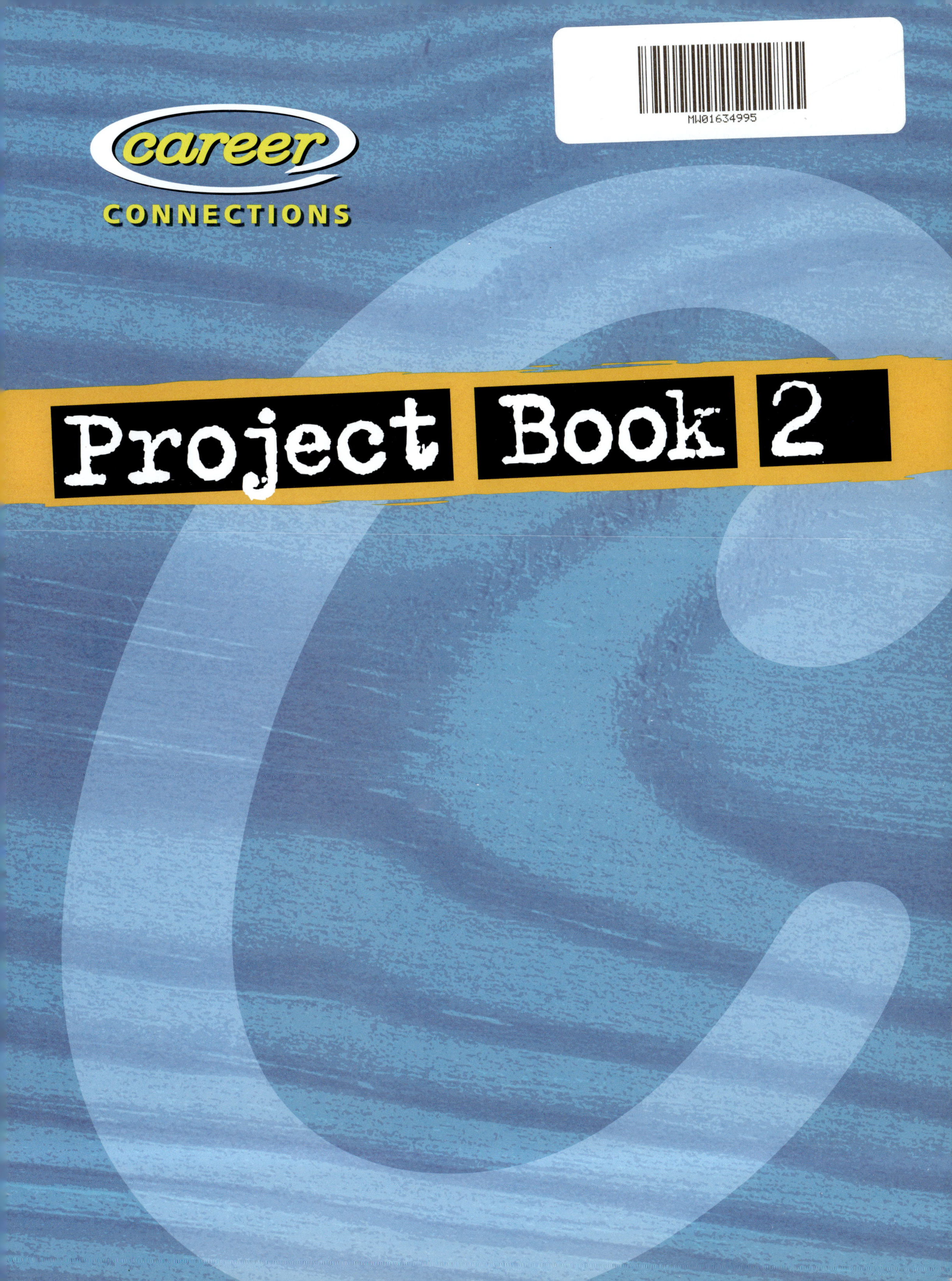
career
CONNECTIONS
Project Book 2

CITF gratefully acknowledges the contributions to the illustration program of Dan Russell, Subject Matter Expert; the staff at 3Plains; the instructors at Chicago Regional Council of Carpenters Training Center and Nelson-Mulligan Carpenters Training Center

Career Connections: Project Book 2 CC0002M
February 2019
ISBN: 978-0-692-20772-7

Send all inquiries to:
Carpenters International Training Fund
212 Carpenters Union Way, Las Vegas, NV 89119-4218
www.carpenters.org

preface

The first priority of the Carpenters International Training Fund (CITF) is state-of-the art training—for the members of the United Brotherhood of Carpenters (UBC) and for the union contractors who employ them. With the *Career Connections* program, CITF brings the skill and experience of its members to high school students and pre-apprenticeship trainees.

The decision-making, goal setting, and employability skills presented in *Career Connections: One Trade, Many Careers* have long been known and are well-documented. Information on carpentry careers is illustrated with actual carpenters at work on jobsites.

The shop projects contained in *Career Connections: Project Book 1*, *Project Book 2*, and *Project Book 3* are designed to develop a progression of skills from basic (*Project Book 1*) to advanced (*Project Book 3*). Lead-Up Exercises allow students to practice skills that will be used in building projects and these Exercises build student confidence in their skills. Step-by-step procedures for each project are supported by drawings and photos illustrating critical steps in the process.

CITF supports this instructional system with the active participation of its instructor-members located across the United States in more than 200 UBC Training Centers. Carpentry programs will have access to these outreach programs for guidance and technical assistance.

A Cautionary Note

The tasks and procedures in this program have been created to teach one or more acceptable and recognized methods of performing specific tasks. These procedures are not meant to be, nor should they be considered, an absolute or complete presentation of the procedures and safety measures related to the tasks.

In the real world of construction, work processes and government regulations can and do change, and it is the employer's responsibility to provide workers with the most recent technical and safety information regarding these processes. Guidelines and instructions presented in this program are not meant to supersede manufacturers' instructions or contractors' jobsite procedures; nor are they meant to replace any current local, state, provincial, or federal safety rules and regulations.

It is essential that everyone always follow all current local, state, provincial, or federal safety rules, regulations, and guidelines in performing any of these tasks.

No statements made in this program should give the impression that the Carpenters International Training Fund or the United Brotherhood of Carpenters and Joiners of America, their affiliates, representatives, or employees have assumed any part of the employer's legal responsibility to provide a "safe and healthful workplace," as mandated by the Occupational Safety and Health Act of 1970.

acknowledgments

The preparation of this program would not have been possible without the technical contributions and cooperative efforts of a wide range of individuals and groups associated with the United Brotherhood of Carpenters (UBC).

Let us start with the members. Without their commitment to excellence, their understanding of the importance of training, and their continued contributions and support, this program would be meaningless. It is their contributions that make this program possible, and this program serves their interests as well as the interests of future workers in the industry.

We would like to acknowledge the instructors and JATCs throughout the United States and Canada who helped develop and review this program. We also recognize the contributions of the members of the Carpenters International Training Advisory Group. Finally, we recognize the ongoing efforts of the staff at the Carpenters International Training Fund (CITF). They have been vital to keeping the development of the program on track and they continue to be key to a successful implementation of the program.

In particular we owe thanks to the following subject matter experts for this program and the Training Programs they represent, which generously allowed them to participate, and the CITF staff who guided and reviewed material.

Thomas Barrett, PhD
Mid-Atlantic Regional Council Joint School of Carpentry, Washington, DC

Paul Bowes
North Texas Carpenters and Millwright Training Center, Arlington, Texas

Lee Libert
CJATC of Greater Pennsylvania, Pittsburgh, Pennsylvania

Michael Maguire
Chicago Regional Council of Carpenters Training Center, Elk Grove, Illinois

Bill McKenna
Washington/Idaho Carpenters Training Program, Kent, Washington

Craig Ramey
Carpenters Training Committee for Northern California, Pleasanton, California

Dan Russell
Washington/Idaho Carpenters Training Program, Kent, Washington

Gary A. Stelzer
Carpenters Joint Apprenticeship Program (CJAP), St. Louis, Missouri

Robert W. Swegle
Mid-Central Illinois Regional Council of Carpenters-JATC, Pekin, Illinois

David Lawson
Senior Technical Coordinator, CITF, Las Vegas, Nevada

Dale Shoemaker
Senior Technical Coordinator, CITF, Las Vegas, Nevada

Jim Vodicka
Technical Coordinator, CITF, Las Vegas, Nevada

Special Thanks to the instructors and apprentices who validated the procedures in this program by building each one of the projects contained in the *Project Books*.

Douglas Lid, *Director;* **Vincent Sticca,** *Assistant Director;* **Michael Maguire,** *Instructor; and the apprentices of the Chicago Regional Council of Carpenters Training Center, Elk Grove, Illinois*

Mark Fuchs, *Program Coordinator;* **Gary Stelzer,** *Retired Coordinator; and the apprentices of Nelson-Mulligan Carpenters Training Center, St. Louis, Missouri*

Louis Russo, *Instructor, and the apprentices of the Southern Nevada Carpenters and Millwrights Training Center, Las Vegas, Nevada*

Vincent Banderman, *Instructor, St. Louis Carpenters Joint Apprenticeship Program, St. Louis, Missouri*

Richard Emerson, *Instructor, Southwest Carpenters Training Center, Reno, Nevada*

James Gault, *Instructor, Detroit Carpenters Joint Apprenticeship School, Detroit, Michigan*

Daniel Hughes, *Instructor, Chicago Regional Council of Carpenters, Chicago, Illinois*

Timothy Moriarty, *Instructor, Connecticut Carpenters Apprentice Training Center, Yalesville, Connecticut*

Project Book 2

contents

to the student

A good life—what does that mean and how do you get it?

A life that has meaningful work, work that makes you proud—what does that mean and how do you get it?

Career Connections: One Trade, Many Careers offers you some ways to understand yourself, and some ways to evaluate all the choices available to you. At the heart of the book is the understanding that you want work that really interests and engages you, work that offers you a variety of experiences, work that offers you opportunities to grow by increasing your skills and earning power.

The Carpenters International Training Fund is pleased to make available to you the experience and skills of its members in the Career Connections program. In all of the books in the Career Connections program—One Trade, Many Careers; Project Book 1; Project Book 2; and Project Book 3—you will get a good idea of life as a skilled and well trained worker: what a jobsite is like, what the working conditions are like, what challenges and satisfactions go along with doing the job. You will have the opportunity to explore yourself—your interests and preferences—and to explore the craft—by learning and practicing the skills you will use as a professional carpenter.

We hope that this program will play a role in bringing you closer to making the connection between your present and your future.

Bill Irwin
Executive Director
Carpenters International Training Fund

what's inside Project Book 2

Career Connections Project Book 2 is organized to provide you with information on all aspects of safety, detailed information on measuring, marking, layout tools; basic carpentry hand tools; power tools for cutting, shaping, fastening, and finishing; and materials and fasteners.

Each section of **the introductory material** gives you an overview of the major topics to be covered in the chapter.

CHAPTER 6

SHED

CONTENTS

238 CAREER CONNECTIONS: PROJECT BOOK 2

Introduction

Your earlier carpentry projects involved mostly construction of picnic tables, outdoor chairs, sawhorses, or workbenches. While working on these worthwhile projects you learned a number of common carpentry techniques. You also developed basic skills in the handling and application of common carpentry tools, materials, and fasteners. In this chapter you will turn your attention to building stand-alone structures. This will enable you to learn more advanced carpentry techniques and skills.

Figure 1
Frame

Stand-alone structures, such as sheds, are built using a frame, as shown in Figure 1. These framing members are like the bones of the house; they give it its shape and structure. Sheets of material called sheathing are nailed over the frame, as shown in Figure 2. The sheathing can be compared to the skin of the house in that it encloses it and keeps it all together. Sheathing also provides a great deal of the strength of the structure.

Figure 2
Sheathing over framing

CHAPTER 6: Shed 239

Importance of Using Tools with Care

Since tools are so helpful and play such an important role in a carpenter's work, they should be shown proper respect. This means understanding how tools work and making a commitment to develop the skills necessary to use them properly. It means never using a tool for any purpose other than the one for which it was designed. It also means keeping tools clean and storing them properly when you are not using them. Above all, it means using tools in a safe manner at all times.

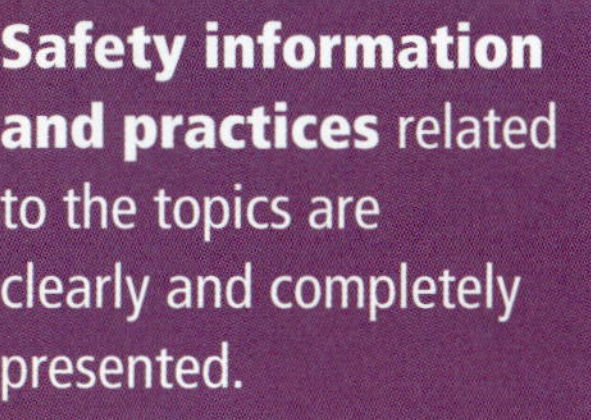

Importance of Tool Safety

No matter how familiar or unthreatening a tool may seem, when used improperly, any tool can cause an injury. That is why students and professional carpenters don't just learn how to use a tool. They learn how to use it safely.

Types of Tools

Carpenters use a broad range of tools. Many of these are hand ...ls, which require only the skilled hands and strength of the ...nter to make them work. Hand saws, block planes, wood ..., utility knives, and tape measures are all examples of ... tools used in carpentry. Other tools used by carpenters ... their power from a source other than human strength— ...tricity for example. Portable circular saws, belt sanders, ...tric drills, and screw guns are examples of power tools. In ...ny cases, a hand tool and a power tool can be used effec...ely for the same task. A skilled carpenter knows how to work ...th both hand and power tools, and understands which tool ... choose for each job.

Safety information and practices related to the topics are clearly and completely presented.

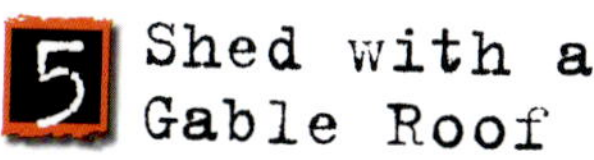

Tools Used for Measuring, Marking, and Layout

A carpenter doesn't just grab a tool and some ... building something. Successful projects take ... Every carpentry project begins as an idea. ... written down in the form of a drawn plan, ... plan may also include a set of instructions. ... start to build, they read the prints or instru... the information they gained from the pri... mark the materials necessary to build the ... of measuring and marking the materials is ...

Prints or instructions contain lengths, ... lines, and angles, all of which must be ca... the materials you intend to use. To transfer ... not only must you be able to read prints, you will also need the appropriate measuring, marking, and layout tools. Using these tools correctly, as well as reading and interpreting prints, takes skill and practice. Layout also requires a basic understanding of mathematics, including geometry.

A **review** of the tools, materials, and fasteners required for the projects reminds you of safety precautions and operating procedures.

...using tools correctly takes skill and practice.

5 Shed with a Gable Roof

The focus of this project is on framing the shed. The framing is basically the same as a professional carpenter would use to build a house, but on a smaller scale. In this project, you will frame a floor, four walls, and a gable roof. A gable roof is a roof that slopes equally in only two directions and has only one type of rafter. You will also make a window opening in one wall and a door opening in another wall. When carpenters build a house, they usually frame the openings for the windows and doors and then install the window and door units into the openings later.

Note: This project has procedures that may require more than one person.

You will need the following materials

Floor System:
- (2) ¾" x 4 x 8 plywood
- (2) 2 x 6 x 8'-0" rim joists
- (4) 2 x 6 x 10'-0" floor joists

Wall system:
- (6) 2 x 4 x 8'-0" top/bottom plates
- (3) 2 x 4 x 10'-0" top/bottom plates
- (36) 2 x 4 x 8'-0" studs/headers
- Scrap ½" plywood for headers

Roof system:
- (1) 2 x 6 x 8'-0" ridge board
- (7) 2 x 4 x 8'-0" rafters

Temporary braces:
- (6) 2 x 4 x 10'-0"

Fasteners:
- 12d sinkers
- 6d common nails

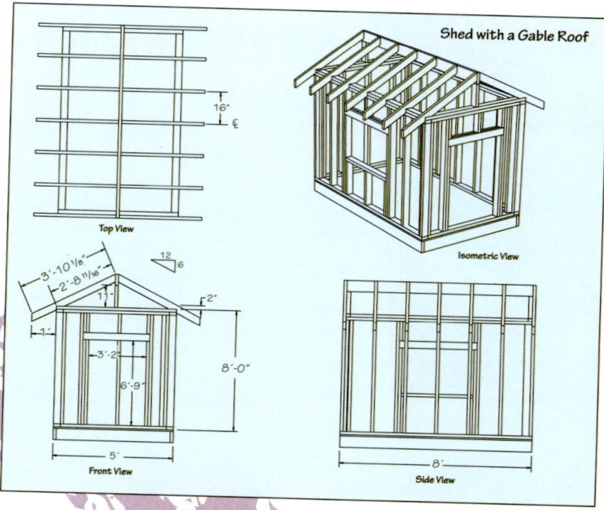

The **presentation of major topics** gives detailed descriptions and illustrations to help you visualize as you learn.

Project Book 2

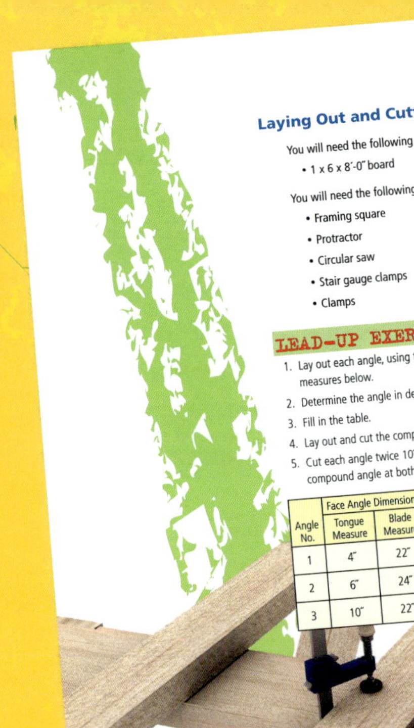

Laying Out and Cutting Compound Angles

You will need the following materials:

- 1 x 6 x 8'-0" board

You will need the following tools:

- Framing square
- Protractor
- Circular saw
- Stair gauge clamps
- Clamps

LEAD-UP EXERCISE

1. Lay out each angle, using the framing square, with the table of measures below.
2. Determine the angle in degrees with a protractor.
3. Fill in the table.
4. Lay out and cut the compound angles from the table below on a 1 x 6 board.
5. Cut each angle twice 10" apart. The result will be 3 boards with the same compound angle at both ends.

Angle No.	Face Angle Dimensions		Angle Degrees	Compound Cut	Bevel Angle Dimensions		Angle Degrees
	Tongue Measure	Blade Measure			Tongue Measure	Blade Measure	
1	4"	22"		A	3"	18"	
2	6"	24"		B	8"	26"	
3	10"	22"		C	12"	12"	

68 CAREER CONNECTIONS: PROJECT BOOK 2

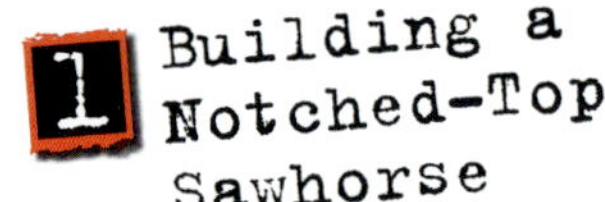

To construct a notched-top sawhorse you will use a variety of measuring, cutting, and fastening tools to cut and assemble components that include both dimensional lumber and plywood. The construction of the sawhorse will be accomplished in three stages or procedures. First you will lay out and cut the legs. Next you will lay out and cut the top, and finally, you will attach the legs, gussets, and side spreaders.

You will need the following materials:

- (1) 1 x 4 x 8'-0" lumber
- (1) 2 x 6 x 4'-0" lumber
- (1) 1 x 4 x 12'-0" lumber
- (1) ½" x 12" x 24" plywood
- (20) 1 ⅝" construction screws
- (16) 1 ¼" construction screws

The construction of the sawhorse will be accomplished in three stages...

CHAPTER 3: Sawhorse 69

The **Materials List** tells you exactly what tools and materials are required to build your project.

The **Lead-Up Exercises** are step-by-step, hands-on practice exercises designed to let you practice the skills you will need to build actual projects.

The **Contents** shows you the three projects from which you and the teacher may choose at every skill level.

CHAPTER 5
SKATEBOARD RAMP

CONTENTS

184 CAREER CONNECTIONS: PROJECT BOOK 2

Introduction

This chapter offers you a choice of three very different and very interesting projects. You may build a skateboard ramp that may very well receive heavy use in your neighborhood, an Adirondack chair that would look good in any garden, or a portable folding workbench that would make a handy addition to any shop. By completing one or another of these projects, you will expand your knowledge of carpentry tools, materials, and techniques. Each of these projects will also give you a chance to practice the carpentry skills you have already learned.

What's New?

The projects in this chapter are intended to expand your carpentry knowledge and skills. Completing a skateboard ramp, Adirondack chair, or portable folding workbench will require you to learn about processes, procedures, tools, and techniques that, as yet, may be unfamiliar to you. Among these are the following:

Caulking Gun A caulking gun is a manual or power tool used to apply caulk or adhesive that is supplied in a tube. See Figure 1. When pressure is applied to the handle or trigger of the caulking gun, it extrudes the caulk or adhesive through a nozzle of the tube.

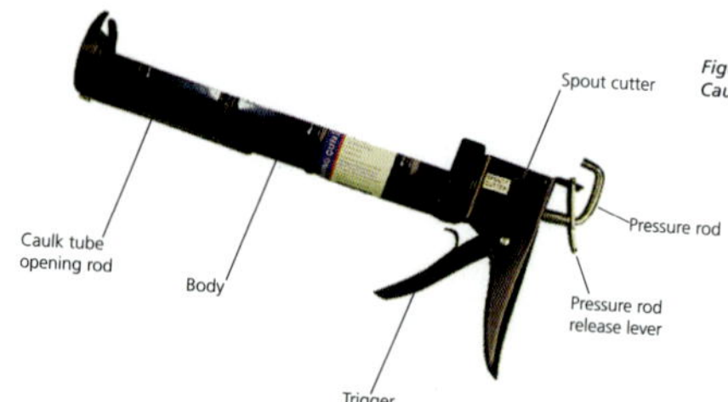

Figure 1 Caulking gun

CHAPTER 5: Skateboard Ramp 185

4. Position one seat slat, part G, at the front bottom of the chair frames and attach it with exterior wood glue and two 2" deck screws. See Figure 61.
5. Attach the two legs, part C, to the front spacer, part J, with glue and 3" deck screws. See Figure 62.
6. Attach the front spacer and leg assembly to the chair frame, part A, with four ⅜" x 4" galvanized carriage bolts with washers. See Figure 63.

Figure 61
Figure 61 illustrates step 4

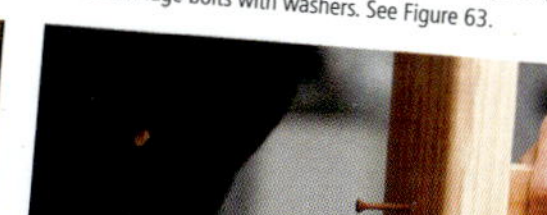

Figure 62
Figure 62 illustrates step 5

Figure 63
Figure 63 illustrates step 6

218 CAREER CONNECTIONS: PROJECT BOOK 2

7. Attach the arm supports, parts F, flush with the top of the legs, part C, and centered left to right with exterior wood glue and three 3" deck screws each through the leg and into the brace. See Figure 64.
8. Place the upper chair back, part I, on the work surface. Scribe a parallel line 1 ⅜" away from the straight edge of the upper chair back along its entire length.
9. Measure and place a mark 3" from each end of the upper chair back, part I, on the scribed line marked in step 8.
10. Draw a line perpendicular to the straightedge of the upper chair back, part I, through each mark made in step 9. See Figure 65.
11. Apply glue to the upper chair back, part I, where the arm will be attached.

Figure 64
Figure 64 illustrates step 7

Figure 65
Figure 65 illustrates step 10

CHAPTER 5: Skateboard Ramp 219

Photographs clearly guide you through the building of your project by illustrating the critical steps.

Figure 90
Figure 90 illustrates step 8

5. Attach the piano hinge to the bottom side of the 2 x 4 x 4'-0" horizontal back face member in the back frame.
6. Lower the tabletop and adjust the side frames until they are square with the back frame.
7. Mark each location where the dowels in the side frames meet the table top.
8. Drill holes ¾" deep at each mark to accept the dowel pins. See Figure 90.

Install the Bottom Shelf

The final stage in construction of the portable folding workbench is to install the bottom shelf. The shelf will be attached to the back rail with strap hinges.

PROCEDURE

1. Mark and install two strap hinges along the plywood bottom shelf, 8" from each end, with ¾" construction screws.
2. Attach the other leaf of the strap hinges to the lower horizontal back rail with ¾" construction screws. See Figure 91.
3. Sand and finish the workbench as desired.

See Figure 91
Figure 91 illustrates step 2

See Figure 92
Completed portable folding workbench

234 CAREER CONNECTIONS: PROJECT BOOK 2

Student Name: ______________________ Date: ____________

Skateboard Ramp Project Evaluation

PROCEDURE	CRITERIA AS SPECIFIED BY THE PROJECT	POSSIBLE POINTS	SCORE
Cutting and Assembly	End cut (2)	5	
	Center support cut	5	
	Blocking cut (16)	32	
	Wedge cut (6)	12	
	Finished wedge assembly (2)	10	
	Blocking installed	5	
	Wedge assembly installed	5	
	Back support cut (3)	7.5	
	Pipe installed	5	
	Plywood back cut and installed	5	
	Ramp and top cut and installed	5	
	Ramp surface installed	5	
	Metal plate installed	5	
	Finished height	5	
	Finished width	5	
	Subtotal	*116.5*	
Overall	Joints tight	5	
	No rough edges	5	
	Screws installed per procedures	5	
	No splits and shiners	5	
	Edges flush	5	
	No excess glue	5	
	Subtotal	*30*	
General	Tool handling	5	
	Followed direction	5	
	Cleaned up	5	
	Safe work practices	5	
	Subtotal	*20*	
		Student score:	

…oints = 166.5

…mum acceptable score: 117 points or ______

…ure: ______________ Teacher's Signature: ______________

CHAPTER 5: Project Evaluations 235

The **Project Evaluation** lets you know exactly how you will be graded and how many points are awarded for each component of your project. The maximum and minimum number of points to achieve a satisfactory performance are shown at the bottom.

before you begin

Career Connections Project Book 2

You are further along on your way to becoming skilled carpenters. It is a craft with many career opportunities. Not many workers are able to pass an office building or family home and say, "I helped build that." The evidence of your work is around you every day and gives you reason to be proud.

The trade of carpentry embraces many crafts. Carpenters often specialize in the interiors of structures; the exteriors of structures; mill cabinet work; and residential carpentry where they work on both the insides and outsides of structures. It is a trade that offers great variety in jobsites and tasks. Work for carpenters is abundant and there are many career opportunities, such as foreperson, superintendent, or even contractor.

As in every other career, the abundance of work and the opportunities depend in a large measure on your skills, focus on the tasks, attention to detail, and the pride you take in doing the work perfectly.

If you are reading this, you have probably completed the projects in *Career Connections: Project Book 1.* Congratulations! The projects in *Career Connections: Project Book 2* are designed to let you practice an intermediate level of carpentry skills building useful objects. You will review the safety practices and attitudes required to keep you and others from injury and the materials and tools from damage. You may need to review the safe and proper operation of some tools and equipment required to build the projects and you will learn how to safely and properly use new tools. An important part of the instruction is given to help you develop a respect for the tools and materials you work with.

The craft of carpentry is an ancient one. It has been practiced and respected for thousands of years. You join a group of skilled and able workers who contribute in a special way to the beauty and usefulness of structures that make everybody's life better.

Enjoy the work you will be doing, the experience of success, and the feelings of pride and satisfaction that come from making something well using your own mental and physical skills. Your next step will be to use advanced skills to prepare you to be a skilled entry-level carpenter.

International Training
Las Vegas

CHAPTER 1

SAFETY ON THE JOBSITE

CONTENTS

What You Now Know

- You are interested in the field of carpentry.
- Safety is important.
- You should not use any tools without being shown how to use them.

What You Will Know

- Safe practices can protect you from injury.
- Common mistakes can lead to accidents.
- You can promote safety and prevent accidents.

1 The Importance of Safety

People tend to believe that accidents and injuries happen to others, but not to them. This is a dangerous way of thinking. The truth is, accidents and severe injuries can happen to anyone, no matter who they are or where they work. The bad news is that accidents are more likely in occupations where tools and powered equipment are used. The good news is that almost every accident can be avoided by simply paying close attention to safety.

You have probably heard the phrase "Safety First." This two-word phrase is very important and carries a lot of meaning. Your safety and the safety of those around you must always be the first priority. This means that safety equipment must always be used or in place when operating any tool or handling any material. It also means that safe work practices must be adhered to at all times when on the jobsite or in the shop area. Safe work practices include:

- Being aware of your surroundings at all times.
- Knowing where and how to exit a building during an emergency.
- Knowing the location of telephones, fire alarms, fire extinguishers, and first aid supplies.
- Making sure electrical connections are safely attached and properly grounded.
- Keeping work areas clean, uncluttered, well lit, and properly ventilated.
- Always wearing clothing and safety equipment that are appropriate for the job you are doing.
- Never attempting to use any tool or handle any material for which you have not been properly trained.
- Choosing the right tool for the job and using it properly.
- Never using a tool for any purpose other than the task for which it was intended.

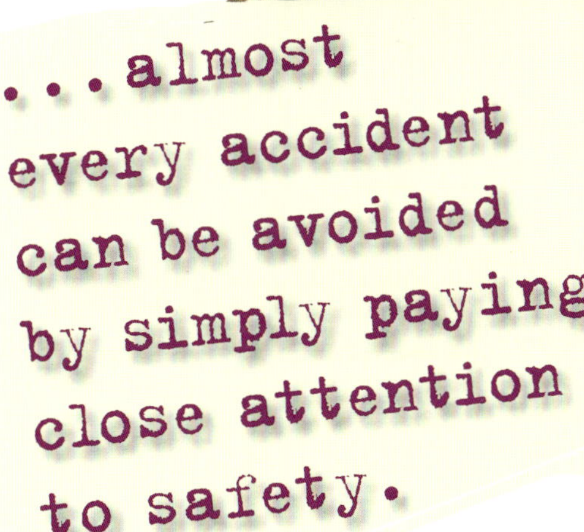

Since carpenters do most of their work with tools, they face a certain amount of risk. For example, tools such as saws have very sharp edges that are used for cutting. When these tools are powered, their cutting edges are moving at very high speeds and will cut anything they come in contact with. Fortunately, the risk of using tools is lessened when the tool is used properly and safe practices are followed. This is why safety must be the first priority whether you have years of experience or are just beginning to learn basic carpentry skills.

2 Hazards, Accidents, and Injuries

Hazards, accidents, and injuries are all closely related. An unrecognized hazard or a hazard that has not been dealt with properly is likely to cause an accident and this, in turn, may lead to an injury. For this reason, recognizing and avoiding hazards are key to preventing accidents and injuries.

What Is a Hazard?

A hazard is anything that may cause harm to you or to others. A hammer left lying on the floor might cause someone to trip, fall, and be injured. The hammer, when left on the floor, becomes a hazard. Using a tool improperly can also be a hazard. For example, failing to keep your hand away from the cutting path of a sharp saw could result in your hand getting cut. In its most simple form, a hazard is an accident waiting to happen.

You are likely to encounter many different hazards in a shop or on the jobsite. It is everyone's responsibility to recognize and eliminate hazards. If you notice a hazard or hazardous situation, you should immediately bring it to the attention of your instructor or supervisor and to others working in the area. That way the hazard can be removed or made less dangerous, and a potential accident or injury can be avoided.

Types of Hazards

Construction work is by its nature a vigorous, physical activity which poses its own set of hazards. These hazards can lead to tripping, falling, or being struck by an object. The injuries that result can include cuts, abrasions, broken bones, and even burns. Hazards that are present on the construction jobsite are also present in the shop or work area. The shop should be treated like any other jobsite, with safety being the prime concern.

The following list includes only a few of the many hazards that may be present at a jobsite or in a shop or work area.

- Tools that have not been properly stored
- Materials that have not been properly stored
- Exposed blades or cutting instruments
- Electrical wires that are not properly attached
- Electrical wires or cables that cross or extend into walkways
- Tools or materials left in walkways
- Dust produced by sawing or sanding
- Improperly stored chemicals, such as paint and solvents, that produce fumes
- Flammable or combustible substances or environments

...recognizing and avoiding hazards are key to preventing accidents and injuries.

What Is an Accident?

An accident is an unplanned event that results in damage to property or injuries to people. Accidents are nearly always unexpected, even so, most accidents in shops and other work areas could have been foreseen and prevented.

Injuries and the accidents that cause them can be prevented by paying proper attention to safety rules...

Accidents nearly always result from hazards that no one has noticed or that have not been dealt with properly. Most accidents are caused by unsafe working conditions or unsafe acts, such as using a damaged electrical cord. By avoiding unsafe behavior and dealing appropriately with work area hazards, such as repairing the damaged cord, most accidents can be prevented.

What Is an Injury?

Hazards cause accidents, and these in turn lead to injuries. An injury is any sort of damage done to the body. An injury may be temporary such as a sprained muscle that will heal, or it may be permanent such as the loss of a limb or an eye. Some injuries can even be fatal.

Ironically, nearly every shop or work area injury could be prevented. Injuries caused by accidents in a work area nearly always point directly to a hazard that was not dealt with properly or to some unsafe behavior that should have been avoided. Injuries and the accidents that cause them can be prevented by paying proper attention to safety rules and practices.

Mistakes that Lead to Accidents and Injuries

The best way to prevent accidents and injuries is to put safety first. Safety must be foremost in your mind when you enter a work area, while you are working, and even when you are done for the day—packing up your tools and storing your materials. There are many specific hazards to watch for when you walk into a shop. The following list includes a few of these hazards.

- Cluttered work area
- Unattended power tools or machinery that are plugged in or left on
- Walkways that are obstructed by tools, materials, or wires
- Spills or slippery spots on floors
- Dust or fumes
- Hazardous or improperly stored materials

Many accidents have more to do with unsafe behaviors, such as inattention, carelessness, and over-confidence...

The hazards listed above may be obvious to a sharp-eyed, safety-minded student or worker, but the causes of many accidents are not so apparent. Many accidents and resulting injuries may have more to do with unsafe behaviors, such as inattention, carelessness, and over-confidence, than with the physical working environment. The following is a list of positive behaviors that may help avoid accidents.

- Proper use of tools
- Proper use of safety equipment
- Safe work habits and attitudes
- Appropriate clothing
- Good communication

Proper Use of Tools

Lack of familiarity with tools is among the most common causes of work-related injuries. You should never use a tool until you have been trained to use it and have demonstrated your ability to operate it properly and safely. Both hand tools and power tools can cause injury when used improperly.

Saws, drills, routers, and other power tools must all be used with caution. Power tools are handy because they enable you to work quickly. Unfortunately, when used improperly, they can cause accidents and injuries just as quickly. Anyone wanting to use a power tool must be completely familiar with it before it is used.

You may think that safety is a less important concern with hand tools than with power tools. This could not be further from the truth. Hand tools require safe and proper handling to avoid injury. An improperly used hand tool has just as much potential of causing serious injury as does a power tool. Proper training and the ability to demonstrate the safe use of a tool are necessary before you operate any tool.

All tools, whether powered or not, must be kept clean and in good working condition. If you notice a tool that is broken, defective, or not working properly, you should report this immediately to your instructor or supervisor. Tools must also be properly stored when not in use. Never hesitate to admit you don't know something, and always feel free to ask your instructor or supervisor for assistance or for more information.

Proper Use of Safety Equipment

You can protect yourself from many different hazards by wearing a hard hat, safety glasses, gloves, sturdy work boots, hearing protection, or other types of safety gear. The safety gear worn by workers is called Personal Protective Equipment (PPE). To protect yourself from injury, you should wear the appropriate PPE at all times when you are working or when you are in a work area.

PPE is not the only safety equipment that should be used. Many power tools and other machinery have their own safety devices, such as guards, that are designed to protect the user. It is important to make sure that these safety devices are in place and functioning properly before using the tool or machinery.

Safe Work Habits and Attitudes

Safety starts with the right state of mind. Almost every accident involves a certain amount of error, and that error is usually in judgment. Carelessness, horseplay, taking unnecessary risks, hurrying, and poor housekeeping can all lead to accidents and injury.

More than any other error, carelessness typically leads to more accidents. It is especially easy to become careless when doing repetitive work. For example, let's say a worker performs a task, such as sawing duplicate lengths of lumber many times a day. That worker may become so accustomed to the repetition of the task that he or she may lose focus on the task itself. In this circumstance, the mind of the worker may wander onto the next task or even to an event that might happen later that night. When focus is lost, accidents and injuries are most likely to happen.

Another danger is a loss of caution after safely operating a tool repeatedly. When a tool or piece of equipment is used for the first time, the person using it tends to be very cautious. Once that person has operated the tool successfully several times, he or she may become complacent. While it is good to feel comfortable and confident with a tool, it is dangerous to lose caution and begin to feel invulnerable, as though an accident won't happen to you. This sense of complacency and invulnerability offers no protection against injuries, including serious ones. In fact, this attitude makes accidents and injuries more likely to happen.

A safe and efficient shop or work area requires everyone to be careful, thoughtful, and professional. Good housekeeping, which is keeping a work space neat and clear of debris, minimizes the risk of injury to you and others. You and your fellow students or workers must be able to count on each other to behave in a safe and responsible manner. Remember that your actions in the shop or work area reveal your personal qualities. Professional carpenters possess the following personal qualities.

- Willingness to take responsibility
- Carefulness
- Thoroughness
- Concern for the safety of others
- Commitment to using tools properly
- Commitment to avoiding all unnecessary risks
- Commitment to completing jobs efficiently and safely

Appropriate Clothing

Many occupations require a specific set of clothing. Police officers and sports figures wear uniforms, people in offices often wear dresses and suits, and firefighters wear protective clothing. This clothing can be used to present a certain image, such as professionalism. It may be designed to allow freedom of movement while engaged in a specialized activity, such as in sports. It may also be chosen for safety purposes, such as the thick coats worn by firefighters. While being a carpenter does not require a specific uniform, it does require proper clothing.

Although a carpenter can wear many different types of clothing, certain types may present a safety hazard. For example, loose fitting clothes or untied shoe laces can cause falls. They can also get caught in equipment and pull a foot, arm, or leg into machinery with disastrous results. It is important to be properly dressed whenever you enter a shop or work area. Clothing should fit. Long hair should be pulled back and tied so that it is out of the way. Items such as rings, bracelets, necklaces, and other loose or dangling jewelry should not be worn while in the shop or work area. Any shoes with laces should be tied. Open-toed shoes offer no protection and should not be worn.

It is important to be properly dressed whenever you enter a shop or work area...

Good Communication

Hazards are far more dangerous when you don't know about them. For instance, you are much more likely to fall and be injured if nobody has told you the floor is slippery. When you see a hazard, you should tell your instructor or supervisor and anyone else in the area. It is also important to listen carefully to warnings passed along to you by your instructors or supervisors and by your fellow students or workers. Be sure you communicate these warnings to anyone who might not have heard them.

4 Preventing Accidents

The best way to avoid an accident is to prevent it from happening in the first place. This can be done if you and others recognize the vital importance of safety and make it a part of your daily work routine. **Being safe and preventing accidents requires the following:**

- Training
- Planning
- Preparation
- Awareness of your surroundings
- Awareness of safety rules
- Good communication
- Good housekeeping
- Skillful and safe use of tools and materials
- Knowledge of hazardous environmental factors
- Safe work habits and attitudes
- Respect for yourself, others, and your shop or work area

Recognizing Hazards

Shops and other work areas are filled with potential hazards. Some of these are obvious while others may be hidden from view. Some hazards may look so ordinary and seemingly harmless that they are easily overlooked. It is essential that every hazard or potential hazard be recognized for the threat it represents to you and others. When determining if something is a hazard, ask the following questions.

- How likely is it that the situation or item will cause an accident or injury?
- What could go wrong if the situation or item is left unchanged?
- What are the likely consequences if it does go wrong?
- What is the source of the potential hazard?

Once you've answered these questions, you will be able to determine if the situation or item is a hazard. You will also have a better idea of the potential hazards you and others are likely to face. If you recognize a hazard or a potential hazard, be sure you tell others about it. Then take steps to eliminate the hazard or make it easier to avoid.

Never use a tool unless you know how to use it properly.

Safe Work Habits

The most effective way to prevent accidents is by developing safe work habits and sticking with them. If you make a habit of using safe working practices, it is less likely that you or someone else will be injured in an accident. Of course, safe work habits don't just happen. They must be developed over time and maintained through proper training, planning, and constant vigilance.

Knowledge and Skill

A safe carpenter is most likely to be a skilled carpenter. Ignorance, lack of skill, or lack of proper training could lead to an accident and injury. Never attempt a task unless you have the knowledge and skill necessary to complete it safely. Never use a tool unless you know how to use it properly. Always feel free to ask your instructor or supervisor for help if there is something you feel you cannot do safely.

Keeping the Work Area Clean

One of the most common causes of accidents is lack of good housekeeping. If you keep the work area clean and organized, you will reduce your exposure to accidents. You must be determined from the very beginning of your project to keep your work area free of debris. By keeping your work area neat from the start, you will demonstrate good housekeeping habits and encourage others to do the same. **The following practices will help make your work area a safer place to learn and work.**

- Clean up spills immediately.
- Position equipment so that wires and cables don't create a tripping hazard.
- Keep walkways and your work area clear of trash and other debris.
- Secure all rugs and mats so that they won't move and their edges won't curl.
- Mark physical hazards with warning signs.
- Wear appropriate shoes and clothing in the work area.

Proper Handling of Materials

Carpentry and other types of construction work often require the moving and positioning of materials. These materials may be quite heavy and handling them can pose a variety of hazards if not done properly. Lifting or carrying a stack of materials that is too heavy for you may injure your back, muscles, or joints. To avoid these injuries observe the following rules.

- Never try to lift or carry anything that is too heavy for you.
- Enlist the help of a fellow student or coworker to move heavy items.
- Whenever possible use mechanical devices such as carts or dollies to help you move heavy materials.
- Position heavy materials as near as possible to the area where they are to be used.
- Use the proper techniques to lift and move heavy loads.

Both wooden and metal materials may have sharp edges and splinters. Avoid sliding your hands along the sides of materials or grabbing the ends too quickly. To avoid injury wear the appropriate gloves when handling materials.

When transporting materials, the whole movement should be planned. It is important to make sure the path is clear and that others in the work area know you are coming. Check the area where the materials will be placed to make sure that it is clear of any potential hazards before moving any materials.

If you are moving large or heavy materials, it's a good idea to ask for assistance. Teamwork is essential when two or more people are carrying heavy or bulky materials. If another person is helping you, it is best if both of you can walk facing forward. Walking backward can be awkward and it is easy to trip and fall. Keep in step with the person assisting you and move at a comfortable pace for both persons.

Call out warnings when approaching obstacles and warn others who may be in the path of the move. Avoid sudden swings and turns that may injure others and jerky, muscle-straining movements. When putting the load down, be aware of where your fingers and feet are in reference to the load. Make sure that both people are in control and slowly lower the material.

Personal Protective Equipment

Use of PPE is required for many tasks. Some PPE, such as eye protection or hearing protection, should be worn whenever you are in a work area. PPE is designed to protect you from certain work-related hazards such as flying debris or loud noises. Personal protective equipment should meet approved industry standards.

- Hard Hat – Usually made of hard plastic, hard hats consist of a tough outer shell separated from the head by a harness. If an object strikes the hard hat, the shell and harness will cushion the blow and protect the head.

PPE won't protect you unless you wear it.

- Eye Protection – The eyes are among the most easily injured parts of the body, and when damage is done to them, it may be impossible to repair. Eye protection allows clear vision while protecting the eyes from flying projectiles such as wood chips. Eye protection also protects eyes from dust and splinters.
- Gloves – For certain tasks protective gloves made of rubber, cotton, and a variety of other materials should be worn. Gloves protect the hands from cuts, abrasions, splinters and exposure to hazardous chemicals. The surface of the glove may be textured or treated with a rubbery substance to provide better grip.
- Protective Footwear – Even in a shop, the toes and feet are exposed to many potential hazards. That's why it is important to wear protective footwear, such as boots. Protective footwear may have reinforced hardened soles to protect the feet from injuries. They may also have metal toe boxes to protect the toes from being crushed by falling objects.
- Hearing Protection – Continuous exposure to loud noises will damage hearing. When working in areas with loud noises, such as the noise from power tools, it is best to wear hearing protection to reduce the noise to safe levels.

PPE won't protect you unless you wear it. PPE should always be used, even when handling a task that may only take a moment. An accident can happen in only an instant, and the injury it causes could affect you for a lifetime.

5 Promoting Safety

Pay close attention to your safety training and encourage others to do so as well.

You are a vital part of the safety and health program at your school or on the jobsite. In fact, a truly effective safety program cannot be maintained without your active participation. You must keep on the watch for hazards, and when you recognize one, point it out immediately to your instructor or supervisor. Pay close attention to your safety training and encourage others to do so as well. And then be sure to put what you learn into practice.

Your school, work area, or jobsite will have a safety plan, and you should learn as much about it as possible. Most safety plans include activities and precautions such as the following.

- Carefully analyze shop or work area for hazards.
- Eliminate or minimize all hazards.
- Report all accidents, injuries, and unsafe conditions to your instructor or supervisor regardless of how insignificant they may seem.
- Make sure everyone knows the location of exits, first aid stations, and emergency supplies.
- Promote safety at all times in the shop or work area.
- Keep tools and equipment clean and in proper working condition.

What to Do in Case of an Accident

Accidents are always unexpected, so you are likely to be shocked when an accident or injury happens to you or someone nearby. Don't lose your head or let your emotions cause you to do something that will make the situation worse. Doing the wrong thing may lead to another accident that may injure you or others. When an accident causes injury to another person, you should:

- Check the area for hazards that may injure you or others.
- Report the accident immediately to your instructor or supervisor and call 911 if necessary.
- Encourage others not to crowd around the injured person as this will make it harder to access the person and provide the needed help.

First Aid Stations

Make sure you know the location of first aid and eyewash stations in your shop or work area. These stations should be well stocked with supplies that are intended to assist you in a variety of safety emergencies. Never remove supplies from a first aid station or an eyewash station for any reason other than to deal with an emergency. Most first aid supplies contain instructions explaining how to use them. You should always follow these instructions even if you think you know how to use the supplies.

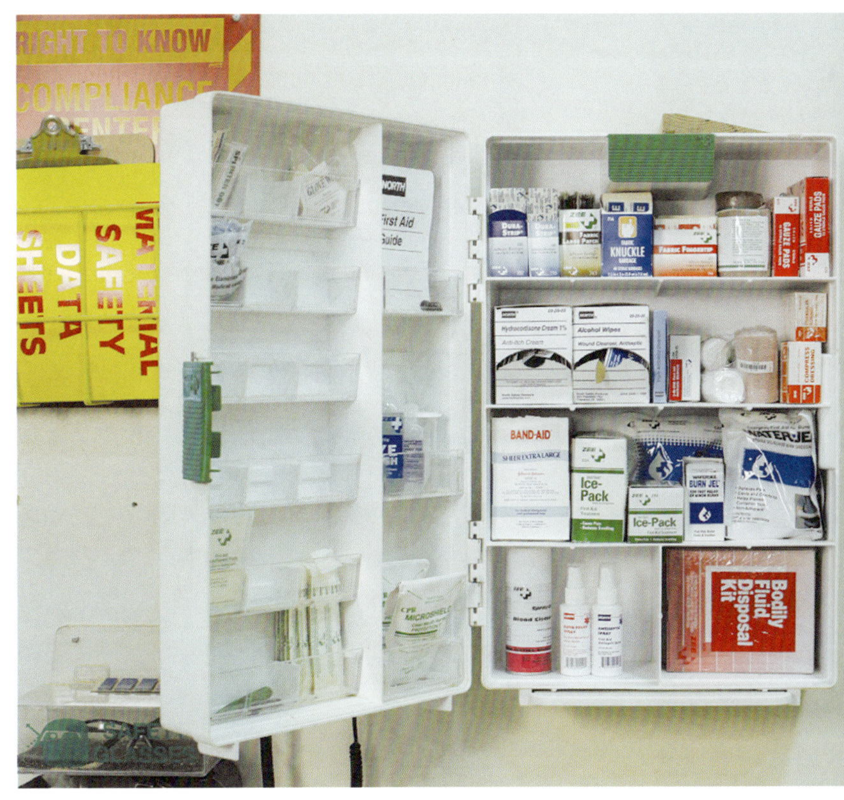

Most first aid supplies contain instructions explaining how to use them.

Chapter Check

On a separate piece of paper, write your answers to the following questions:

Multiple Choice

1. Which of the following is defined as an unplanned event that results in damage to property?
   a) hazard
   b) injury
   c) accident
   d) violation

2. Which of the following can often result in carelessness when operating a tool?
   a) using a tool repeatedly
   b) wearing gloves
   c) using the wrong tool
   d) awareness of your surroundings

3. Which of the following is a form of PPE?
   a) flammable substance
   b) abrasion
   c) good communication
   d) safety glasses

4. Which of the following is an example of a safety device on a tool?
   a. gloves
   b. hazard
   c. PPE
   d. guards

5. Keeping your work area clean is known as which of the following?
   a. personal protective equipment
   b. good housekeeping
   c. ergonomics
   d. good communication

Fill in the Blank

6. ____________________ is an important two-word phrase that means safe work practices must always be your main priority.

7. Anything that can cause harm to you or to others is known as a(n) ____________________.

8. Lack of ____________________ with tools is one of the most common causes of work-related injuries.

9. You should know the location of the ____________________ in your work area in case you need emergency safety supplies.

10. A(n) ____________________ is any sort of damage done to the body.

True or False

11. Safety is a less important concern with hand tools than it is with power tools. *T or F*

12. Storing tools properly is a way of practicing safe work habits. *T or F*

13. Loose clothing is considered a safety hazard. *T or F*

14. Lack of familiarity with tools is among the most common causes of work-related injuries. *T or F*

15. When you see a hazard you should tell your instructor, supervisor, or anyone else in the area. *T or F*

CHAPTER 2

REVIEW OF TOOLS, MATERIALS, AND FASTENERS

What You Now Know

- Tools make it easier to accomplish a task.
- Safety is an important part of tool use.

What You Will Know

- The importance of tools to a carpenter
- How to use measuring, marking, and layout tools

CONTENTS

1 Handling Tools with Care

In *Career Connections: Project Book 1* you were introduced to a variety of the tools and materials used by professional carpenters. You used both hand and power tools as well as several kinds of fasteners to shape wood and other materials into useful products. In this book, *you* will increase your experience with these tools and materials while practicing your carpentry skills. But before you go on, take the time to read this chapter in order to refresh your knowledge about the tools and materials of the trade.

Very little work of any kind could be accomplished without the assistance of tools. This is especially true for carpenters who must rely on a wide variety of hand tools and power tools to get their work done. However, carpenters understand that their tools are effective only when used with skill and care. Improperly used tools can destroy expensive materials, damage the products carpenters are trying to create, and cause serious injuries.

Among the most important skills carpenters must learn is the ability to use their tools safely and efficiently. This requires a commitment to treat tools and materials with respect and to observe safety precautions at all times. Experienced carpenters observe the following rules for handling and maintaining tools.

Know Your Tools Never use any tool without proper supervision unless you are completely familiar with it and skilled in its use. Lack of familiarity and skill with tools is a major cause of work-related injuries.

Keep Tools Clean Oil, dust, dirt, and grime will interfere with the effectiveness of a tool and make it harder to control safely. For these and other reasons, tools should be kept clean at all times.

Keep Tools in Good Repair A damaged tool is unlikely to perform its task effectively. It may damage or destroy materials or injure the carpenter attempting to use it. This is why it is very important to inspect tools for damage or defects before you use them. If any tool shows signs of damage or isn't working properly, it should be taken out of service immediately and repaired or replaced. A tag which identifies the reason it was taken out of service should also be placed on the tool or piece of equipment, as shown in Figure 1.

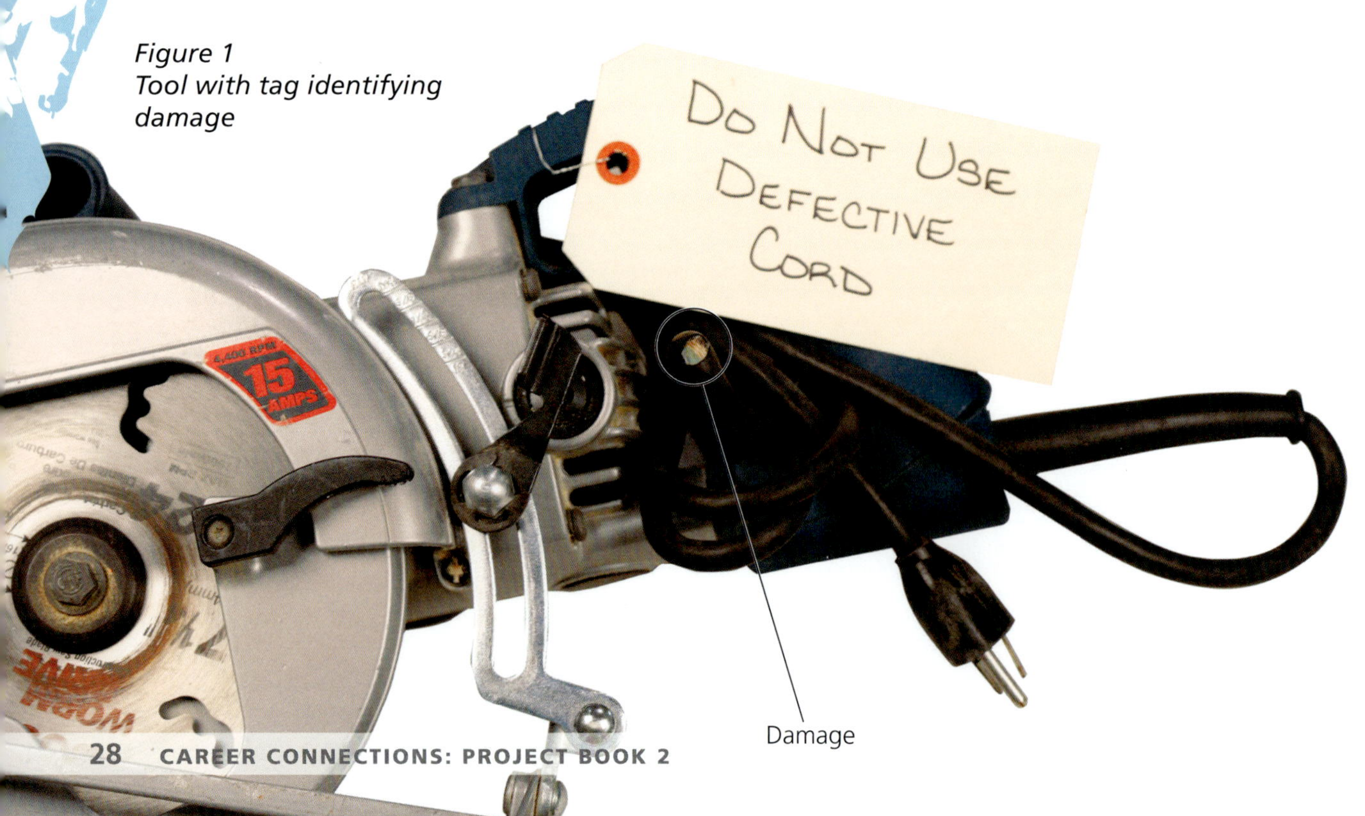

Figure 1
Tool with tag identifying damage

Properly Store Tools When Not in Use To protect tools and avoid clutter that could cause accidents or injuries, tools must be properly stored when not in use. Blades should be covered to keep them sharp and prevent accidental cuts. As shown in Figure 2, cords should also be coiled properly so they don't get tangled and possibly create a tripping hazard.

Keep Hands Away from Cutting Edges Keep hands and other body parts away from the blades and sharp edges of saws, knives, chisels, and other cutting and shaping tools. Always stand behind cutting and shaping tools when they are in use to prevent being struck by kickback or flying debris.

Keep Power Cords Out of Walkways Power tools are often driven by electricity drawn from wall outlets. Improperly placed power cords may represent a dangerous tripping hazard. It is also very important to keep power cords out of the path of cutting and shaping tools and away from sharp blades.

Take Precautions Against Electric Shock Make sure the power cord you are using is properly grounded with a three-pronged plug or that the power tool itself is double insulated. All power cords should be inspected for frays, exposed wires such as the cord in Figure 1, breaks in the insulation, or wires pulled out of the plug or handle.

Wear Proper Protective Gear Always wear the appropriate personal protective equipment (PPE) when using tools or when entering an area where tools are in use.

Guard Against Accidental Startups Before changing the blade or settings of a power tool, make sure it is turned off and disconnected from its power supply.

Figure 2
Cords coiled properly

2 Hand Tools

In *Career Connections: Project Book 1*, you were introduced to more than a dozen common carpentry hand tools. You will be using many of these same tools again while completing the projects outlined in this book. These tools are listed below along with a brief review of each.

Block Plane

Planes are shaping tools used to smooth the surface of wood or to remove material from components so that they fit together properly. This is often done by hand using a small block plane, like the one shown in Figure 3. A block plane consists of a metal body and an adjustable metal blade, called a plane iron, which fits into a recess in the body. Designed to be used with one hand, most block planes have fingertip rests and a rounded lever cap that fits into the palm of the hand. The blade of a block plane is held at an angle of 20° to 25° with the bevel of the plane iron pointing upward.

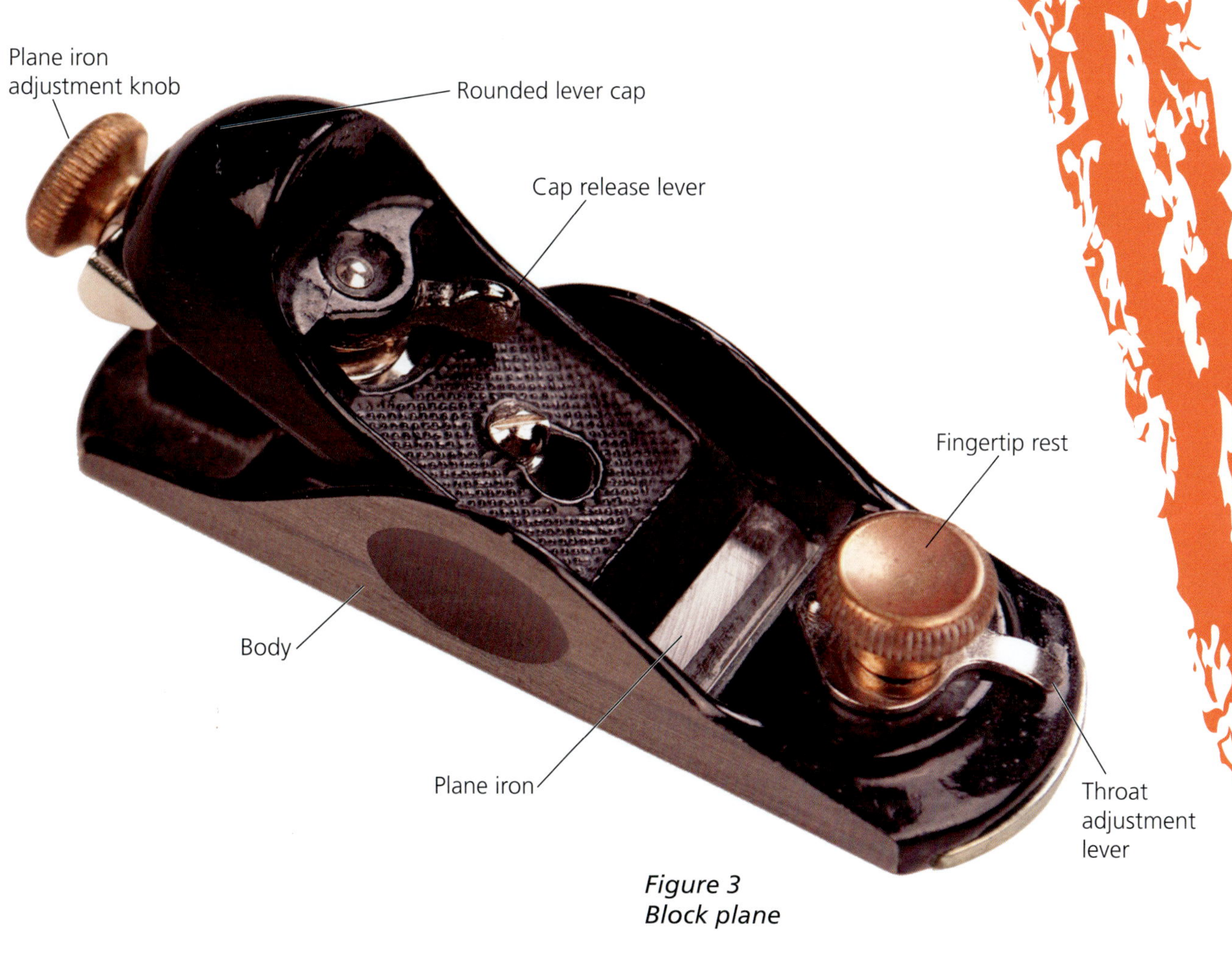

Figure 3
Block plane

Chalk Box

Carpenters often use a chalk box for marking long straight lines. A chalk box, like the one shown in Figure 4, consists of a string wound inside a container filled with powdered chalk. When the string is pulled out of the container, it is covered in chalk. The string is then stretched between two marks and pulled upward slightly. When the string is released it will leave a distinct straight line on the surface below, such as the one shown in Figure 5. Once the line has been snapped, the string is rewound into the chalk box by turning a handle on the side of the container.

...use a chalk box for marking long straight lines.

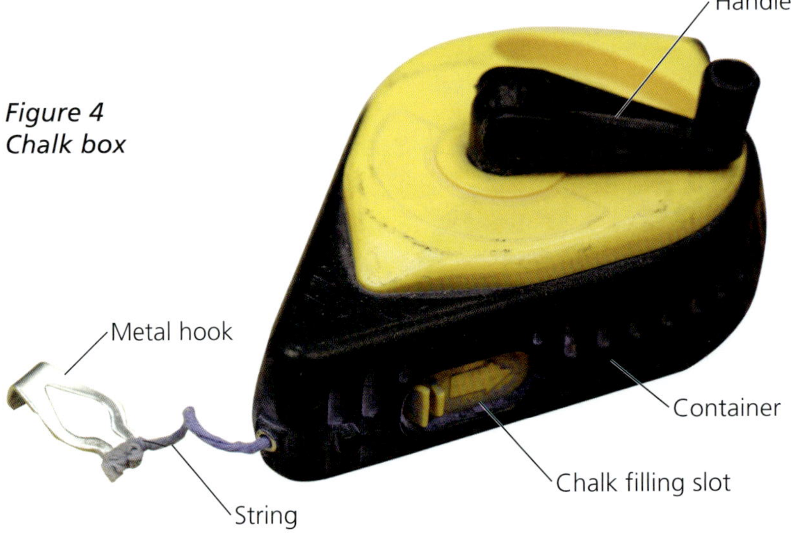

Figure 4
Chalk box

Figure 5
Chalk line

Chisel

An extremely versatile tool used for cutting, trimming, fitting, and shaping wood, a chisel consists of a handle and a metal blade often referred to as an iron. An example of a chisel is shown in Figure 6. In most cases, the chisel blade is a flat steel bar beveled on one end at an angle of 25° to 30°. The beveled edge is sharpened so that it will cut or shave wood when the chisel is struck with a mallet, hammer or manipulated by hand.

Chisels must be used with caution to avoid severe cuts and other serious injuries. When using a chisel, keep both hands behind the cutting edge and the chisel pointed away from the body. Never place a chisel in a shirt or pants pocket since the sharp edge may cut through the material and cause an injury. When a chisel is not in use, cover the blade with a protective cap like the one shown in Figure 7.

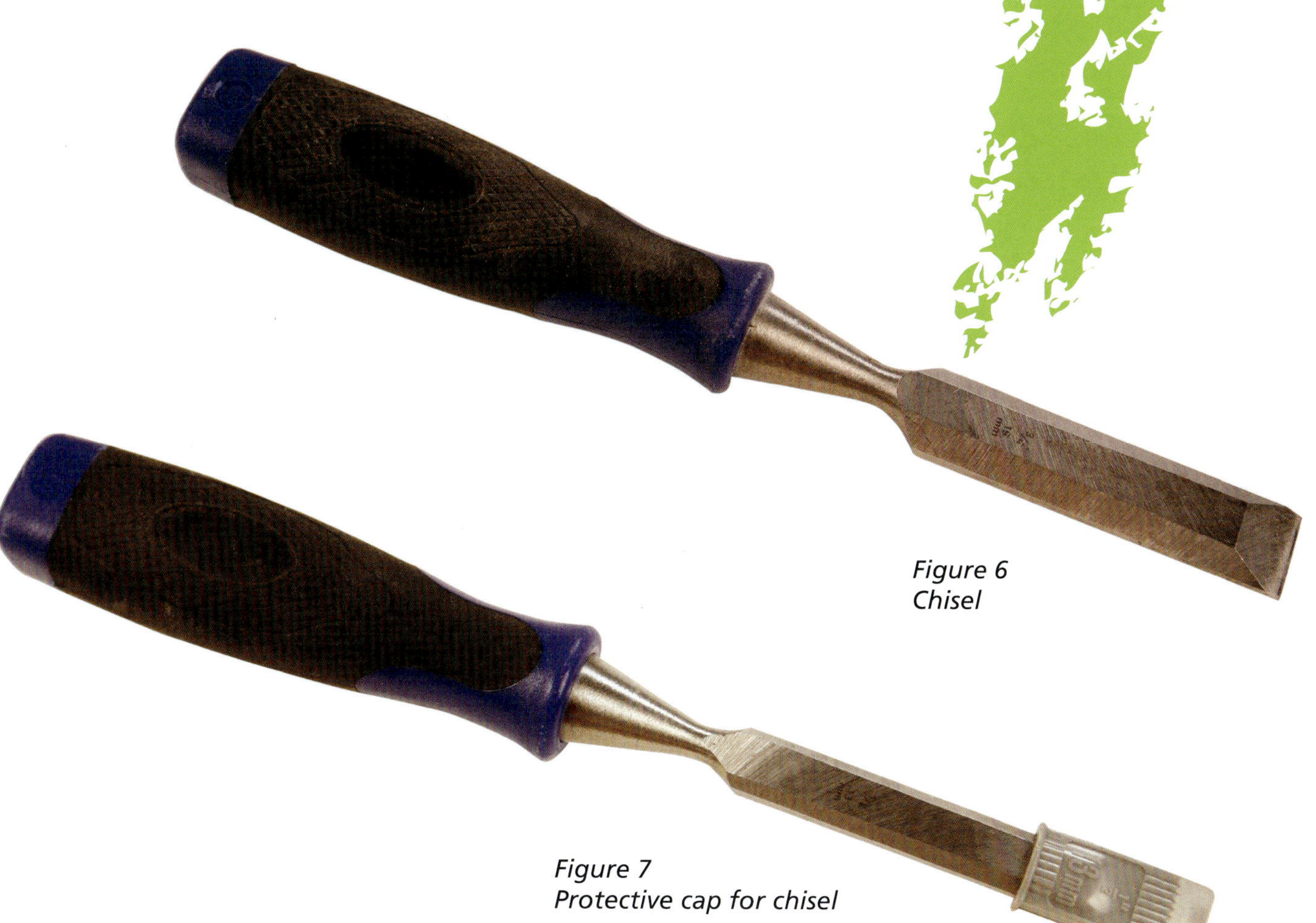

Figure 6
Chisel

Figure 7
Protective cap for chisel

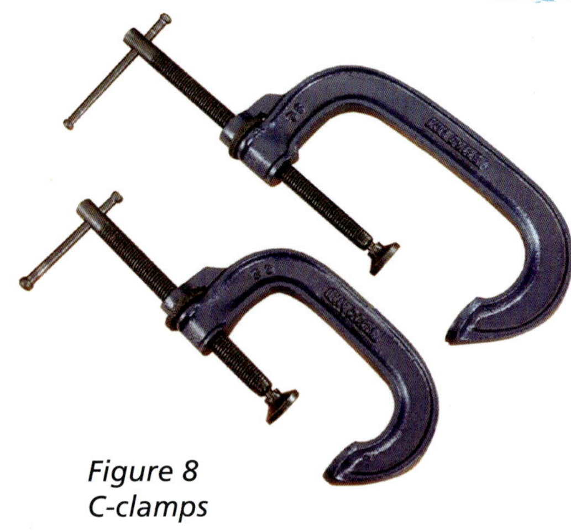

Figure 8
C-clamps

C-Clamp

Like all clamps, C-clamps are used to hold one or more components or pieces of material in place while they are being worked or while an adhesive sets. Shown in Figure 8, C-clamps are shaped like the letter C. The material the carpenter wants to secure fits into the open end of the C-clamp where it is held in place by a metal shaft with threads like those on a bolt. The shaft is twisted or tightened until it presses the material firmly against the other end of the clamp.

Bar and Pipe Clamps

Carpentry materials are sometimes held in place with either bar or pipe clamps. Each of these clamps consists of a long metal piece with plates or jaws. With a bar clamp the metal piece is rectangular as shown in Figure 9A. With the pipe clamp it is round, as shown in Figure 9B. One jaw on both clamps is usually adjustable while the other is securely fixed to the end of the bar or pipe. The adjustable jaw slides along the length of the bar or pipe and pushes against the material held in the clamp. The jaws are tightened either by a powerful spring or a screw shaft with a handle.

Figure 9
(A) Bar clamp

Figure 9
(B) Pipe clamp

Hand Screw Clamp

A hand screw clamp, as shown in Figure 10, consists of two wood jaws connected by a pair of threaded rods. The wood jaws are usually made of hard maple, and the threaded rods are made of steel. The rods are fitted with handles, and the clamp is tightened by twisting them. Since the rods can be tightened independently, it is possible to use these clamps on angled surfaces. A hand screw clamp can be used like a vice to hold material upright.

Figure 10
Hand screw clamp

Claw Hammer

A claw hammer, like the one shown in Figure 11, is a striking tool consisting of a handle and a heavy metal head. Handles are made of wood, plastic, fiberglass, or steel. Usually made of drop-forged tempered steel or titanium, hammer heads generally consist of a face for striking or driving, flat sides called cheeks, and claws for pulling nails. The claws can be either curved or straight. Curved claws provide extra leverage for removing nails.

Because hammers have heavy metal heads, they can cause serious injury when used improperly. When you have a hammer in your hand, always keep a firm grip on the handle. Do not strike a nail or any other object with the cheek of the hammer and never use claw hammers to strike other hardened steel surfaces.

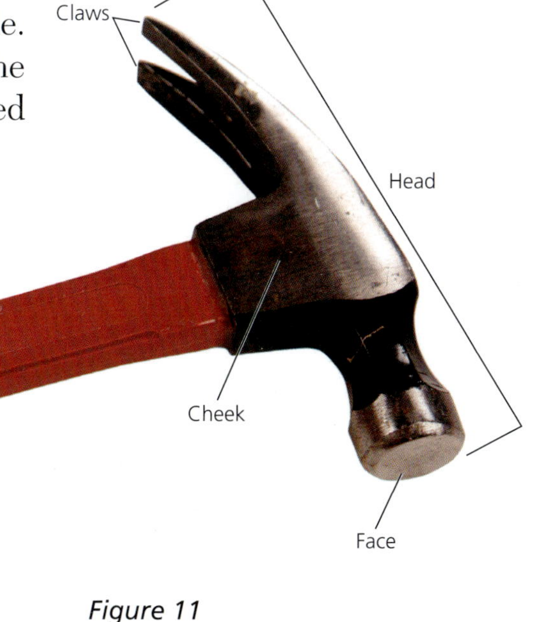

Figure 11
Claw hammer

Nail Set

A nail set is a round, tapered metal shaft, usually about 3 ½″ long, which has a cupped tip designed to fit on top of a brad or finishing nail. A nail set is shown in Figure 12. For appearance's sake, carpenters often fasten components with thin brads or finishing nails, and then use a nail set and a hammer to drive them below the surface of the material where they are less likely to be noticed. The nail set is held against the head directly in line with the brad or nail and then struck lightly with a hammer until the head is set to a depth roughly equal to the brad or nail's diameter.

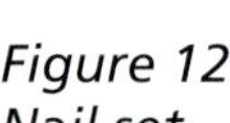

Figure 12
Nail set

Combination Square

Used by carpenters for a wide variety of measuring, marking, and layout tasks, a combination square is a highly versatile tool. Shown in Figure 13, a combination square consists of a handle and an adjustable blade marked off on both sides in inches and fractions of an inch. The head is designed with both a 45° and 90° angle in relationship to the blade. The blade is grooved so that it can be held in place with the locking device. A locking device in the handle can be tightened to fix the blade in place at a desired measurement. Many combination squares are fitted with built-in levels that can be used to determine whether the handle is exactly horizontal and the blade is exactly vertical. Some combination squares have a removable marking device built into the handle.

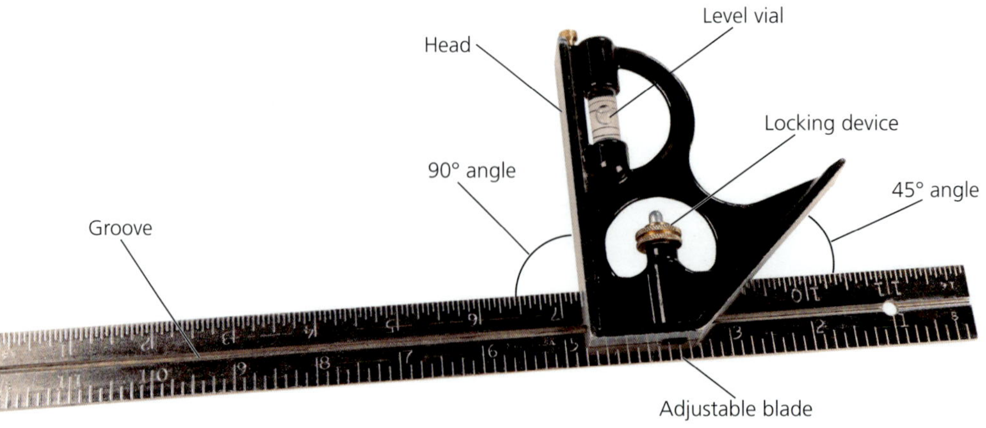

Figure 13
Combination square

Compass

Used to lay out circles and arcs, a compass has two metal legs hinged at one end so that they can be spread apart and locked into place at any desired width. As shown in Figure 14, one leg usually has a sharp point at the end while the other leg is designed to hold a pencil. Circles and arcs are drawn by placing the point of the compass on paper or on a piece of carpentry material and then rotating the pencil around the point. The sharp metal point of a compass can cause serious injury if you or others are jabbed with it, so this precision instrument should be handled with care.

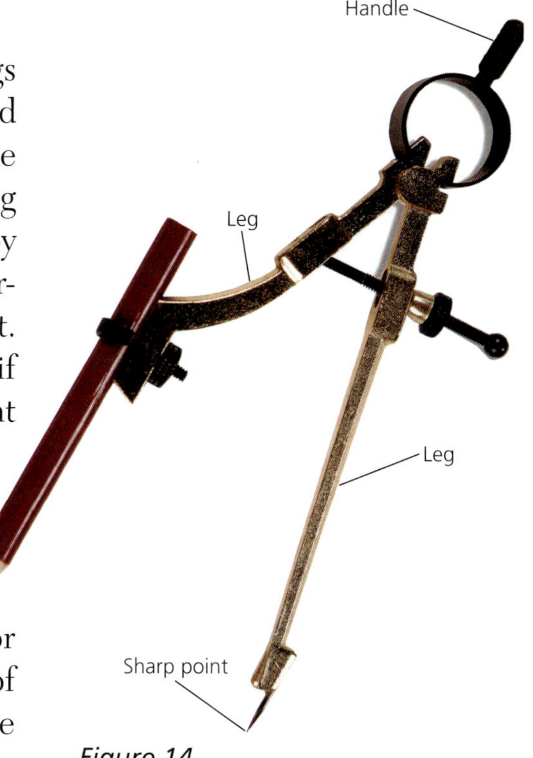

Figure 14
Compass

Framing Square

One of several measuring and layout tools frequently used for marking and checking 90º angles, a framing square consists of a 24″ metal body and a 16″ metal tongue arranged in the shape of an L. A framing square is shown in Figure 15. Usually, the body of the square is 2″ wide and its tongue is 1 ½″ wide. In most cases, both sides of a framing square are marked with useful scales and tables, such as a rafter table. A framing square can become damaged and must be checked often to make sure the angle formed by the body and tongue is exactly 90º.

Figure 15
Framing square

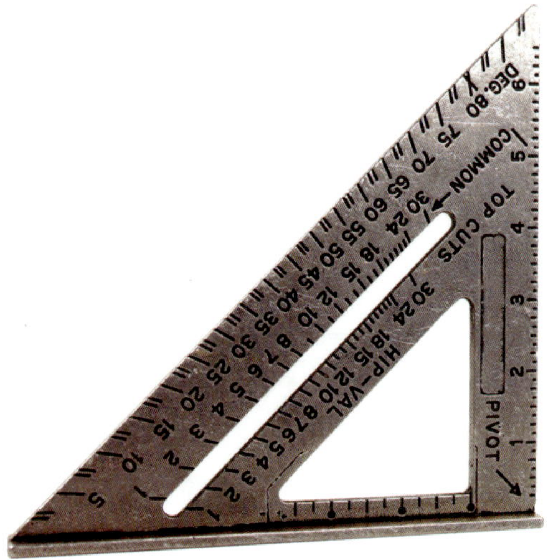

Figure 16
Speed® Square

Speed® Square

Carpenters use a number of different tools to check for square. Among the most convenient and easiest to use of these is the Speed® Square. A Speed® Square is a flat, three-edged device with a lipped edge and a 90° angle at one corner. The Speed® Square is used not only to check right angles but also for laying out a variety of other angles and for drawing straight lines. The speed square's small size makes it easy to store and to carry around in a work area. Speed® Squares come in two sizes, a 7″ model and a 12″ version. The 7″ version is shown in Figure 16.

Straightedge

Carpentry layouts consist largely of straight lines. Whether laying out projects or cutting materials, carpenters must make sure these lines are as straight as possible. A metal yardstick, shown in Figure 17, or even the factory edge of a piece of plywood may be used to help a carpenter draw or cut straight lines. Anything may serve this purpose so long as its edge is smooth and forms a perfectly straight line.

Figure 17
Straightedge

Retractable Tape Measure

Figure 18
Tape measure

Compact and easy to carry, tape measures provide carpenters with a quick and convenient way to determine the dimensions of an object. A retractable tape measure, like the one in Figure 18, consists of a long, flexible strip of metal marked off in inches and fractions of an inch or in metric units. Usually, the tape or blade is housed in a durable, spring loaded case and can be pulled out to the required length. Once the necessary measurements have been taken, the blade can be retracted into its protective case.

In America most tape measure blades are marked off in feet and inches and in fractional increments of 1/2″, 1/4″, 1/8″, and 1/16″. Some also have markings accurate to within 1/32″. Tape measure blades are fitted with a hook at the end designed to fit over the edge of an object when taking outside measurements. These hooks are designed to slip inward slightly to compensate for the thickness of the hook when taking inside measurements.

Protractor

Carpenters often use protractors to measure or mark angles and arcs when doing a layout. Made of either metal or plastic, most protractors are semicircular in shape. Degree measurements are marked along the curved outer edge of the protractor ranging from 0° on the far right to 180° on the left. Along the inside curved edge, degrees are marked from 0° on the far left to 180° on the far right. Some protractors are circular and are marked from 0° to 360°. Both types are shown in Figure 19.

Figure 19
Protractors

Sliding T-Bevel

A sliding T-bevel is an adjustable tool used to lay out angles or transfer angles. Shown in Figure 20, it consists of a slotted metal blade attached to a handle made of either wood or metal. The blade is attached to the handle with a bolt and a wing nut. The slotted blade is slid through the handle until it reaches the desired length or angle and then is locked in place by tightening the wing nut or locking device. The angle of the blade can be set using a protractor.

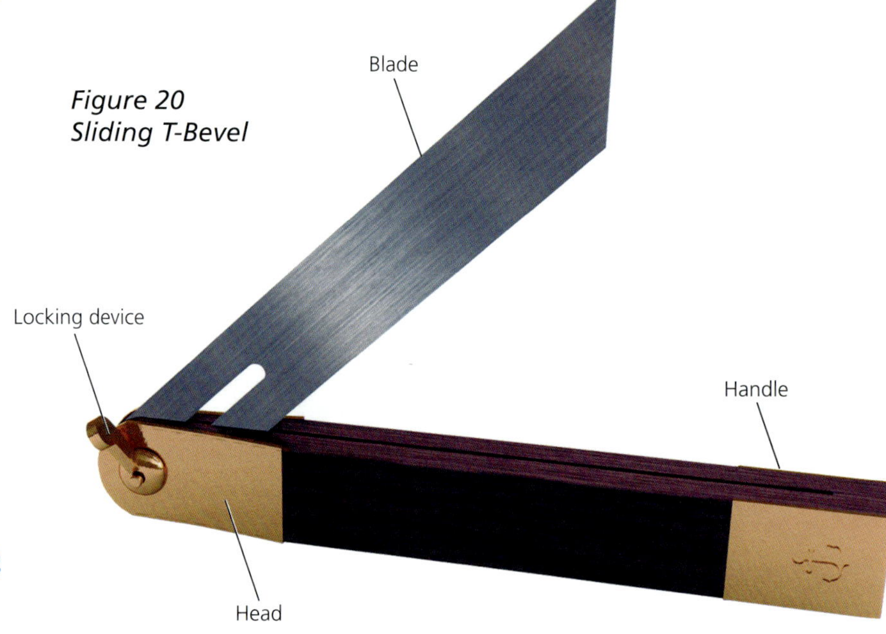

Figure 20
Sliding T-Bevel

The jaw width of an adjustable wrench can be easily changed...

Adjustable Wrench

The jaw width of an adjustable wrench can be easily adjusted to enable the wrench head to fit many different sizes of nuts and bolts. Usually, the width is adjusted by turning a roller that moves one of the jaws. An example of an adjustable wrench is shown in Figure 21.

Figure 21
Adjustable wrench

Socket Wrench

Socket wrenches, like the one shown in Figure 22, come with interchangeable cylinder-shaped sockets designed to fit over many different sizes of nuts and bolts. The wrench handle and socket are linked by a ratchet, which is a mechanism that engages when the handle is moved in one direction and disengages when the handle is moved in the other direction. The ratchet makes it possible to quickly tighten or loosen a nut or bolt without repositioning the socket. A switch on the back of the ratchet reverses the direction of the ratchet gear so that the wrench can be set for either tightening or loosening.

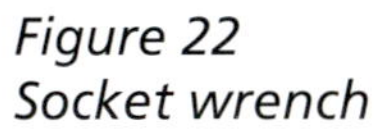

Figure 22
Socket wrench

Open End and Box End Wrenches

Consisting of a length of tempered steel with a head at one or both ends, an open end and box wrench is used for tightening or loosening nuts and bolts of various sizes. The head of the wrench is placed over the nut or bolt and pressure is placed on the opposite end of the handle. The head of the wrench must be the correct size or it will not fit the nut or bolt. Open end and box end wrenches, also called combination wrenches, are shown in Figure 23.

Figure 23
Open end and box end wrenches

Wood Rasp

A rasp, shown in Figure 24, is a shaping tool similar to a file. The metal blade of a rasp is crisscrossed with coarse ridges, or teeth, that cut into wood as they are pushed across a piece of lumber or wood material. The teeth of a rasp cut deep and remove more material with each pass than can be removed with a file. This makes a rasp ideal for rough shaping wood but less useful for smoothing surfaces.

Figure 24
Wood rasp

Sanding Block

Designed to be used with sandpaper or some other abrasive material, sanding blocks are used when manually smoothing a surface. Usually, the size and shape of a sanding block makes it easy to grip it firmly in one hand. An example of a sanding block is shown in Figure 25.

Figure 25
Sanding block

Utility Knife

Utility knives, like the ones shown in Figure 26, are highly portable and provide a convenient way to handle a wide variety of light cutting tasks. Carpenters often use utility knives for scoring, scribing, shaving, and shaping materials as well as for cutting. A utility knife consists of a metal or plastic handle with either a fixed or a retractable blade. Utility knife blades are razor-sharp when new, but they will lose their edge quickly with heavy use. Most utility knives are equipped with a storage space for spare blades.

When used improperly, utility knives can be very dangerous. To avoid severe cuts and injuries, they must be used with extreme care. It is very important to keep hands and fingers away from the cutting edge at all times.

Figure 26
Utility knives

3 Power Tools

Driven by electric motors or by compressed air, power tools are faster and more efficient than hand tools, and they greatly reduce the need for heavy hand labor. Although not discussed in this section, power tools may be run by gasoline motors or may be powder-actuated. Nearly every job that was once done with hand tools can now be done more quickly using power tools. There are power tools available for routing, sawing, drilling, nailing, stapling, planing, sanding, and many other processes. Although extremely useful, power tools can also be very dangerous. When working with power tools, it is important to take all the appropriate precautions and to follow the safety rules specific to each power tool.

In *Career Connections: Project Book 1*, you were introduced to a number of power tools often used by professional carpenters. You will be using many of these same power tools again while completing the projects outlined later in this book. These power tools are shown on the following pages along with a brief review of each.

...power tools are faster and more efficient than hand tools...

Belt Sander

A belt sander, like the one shown in Figure 27, is a handheld finishing tool that moves a continuous abrasive belt over the surface of a material. Belt sanders are especially helpful when large surfaces must be finished or a large amount of material must be removed. Available in many different designs and models, belt sanders often have heavy-duty motors and two handles to make them easier to control. Because they generate a large amount of dust, belt sanders should be used with a dust collection system whenever possible.

Figure 27
Belt sander

Portable Circular Saw

A portable circular saw, which is usually just called a circular saw, is a cutting tool with an electric motor that rotates a disc-shaped blade at a high rate of speed. An example is shown in Figure 28. The blade of a circular saw will quickly slice through lumber, plywood, and many other materials. For safety, hand-held circular saws have blade guards that cover both the top and bottom of the blade. The upper portion of the blade is covered by a fixed guard, while the bottom portion is covered by a spring-loaded guard that rotates backward as the saw is making a cut. This exposes only the portion of the blade that is cutting into the material. When the saw is removed from the material, the bottom guard springs back into position to cover the blade. Depending on the type of cut required, both the cutting depth and cutting angle of circular saw can be adjusted.

Switch
Second handle
Handle
Tilt adjustment
Power lead
Upper blade guard
Lower guard lifting handle
Base
Circular saw blade
Bottom blade guard

Figure 28
Circular saw

Safety is an especially important consideration when using a circular saw. As with other power tools, you should not attempt to use a circular saw without supervision unless you have the proper training. Some of the safety concerns when using a circular saw are as follows.

- Before using a circular saw, inspect the cord and plug for cuts, frays, or a missing ground prong.
- Make sure the material you are cutting is properly supported so that it will not bind the blade.
- Be sure the retractable guard on the blade is in place and operational before plugging in the saw.
- Always unplug the saw before making any adjustments or changing the blade.
- Always keep a firm grip on a circular saw while it is in use and never put a circular saw down until the blade has completely stopped turning.

...safety is an especially important consideration when using a circular saw...

Miter Saw

Specially designed for cutting at an angle, miter saws are used to conveniently produce miter cuts, bevel cuts, compound miter cuts, and a variety of other angled cuts. There are several types of miter saws, but the most versatile is the dual-bevel sliding compound miter saw like the one shown in Figure 29. The sliding compound miter saw can be tilted left or right to make precise cuts at many different angles.

Like other cutting tools, sliding compound miter saws are dangerous and can cause injuries if not used properly. They should never be operated without supervision by anyone who lacks the necessary experience. The safety precautions that should be followed when using a sliding compound miter saw are very similar to those required for using a circular saw.

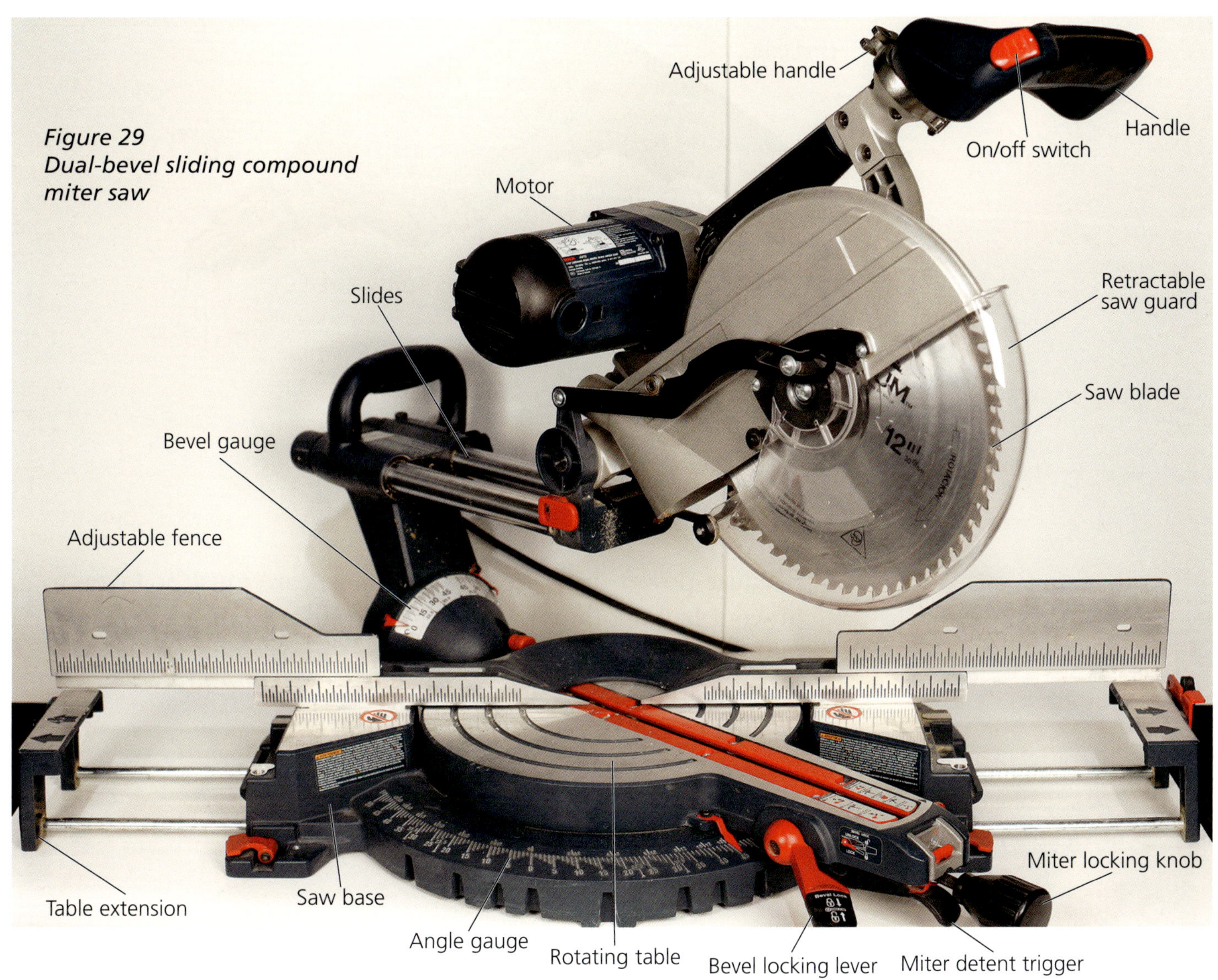

Figure 29
Dual-bevel sliding compound miter saw

Electric Drill

A relatively simple tool, an electric drill consists primarily of an electric motor with an attached handle and a chuck for holding various bits. An example of an electric drill is shown in Figure 30. The chuck is the mechanism that allows drill bits to be attached to the drill motor. Common drill chuck sizes are 1/4″, 3/8″, or 1/2″. The size refers to how large a drill bit shank the chuck will accept. The most common chuck size is 3/8″. Power drills can be used with a variety of drill bits to bore holes through many different types of material with relative ease. Power drills can also be used with special discs or drums for sanding and even polishing.

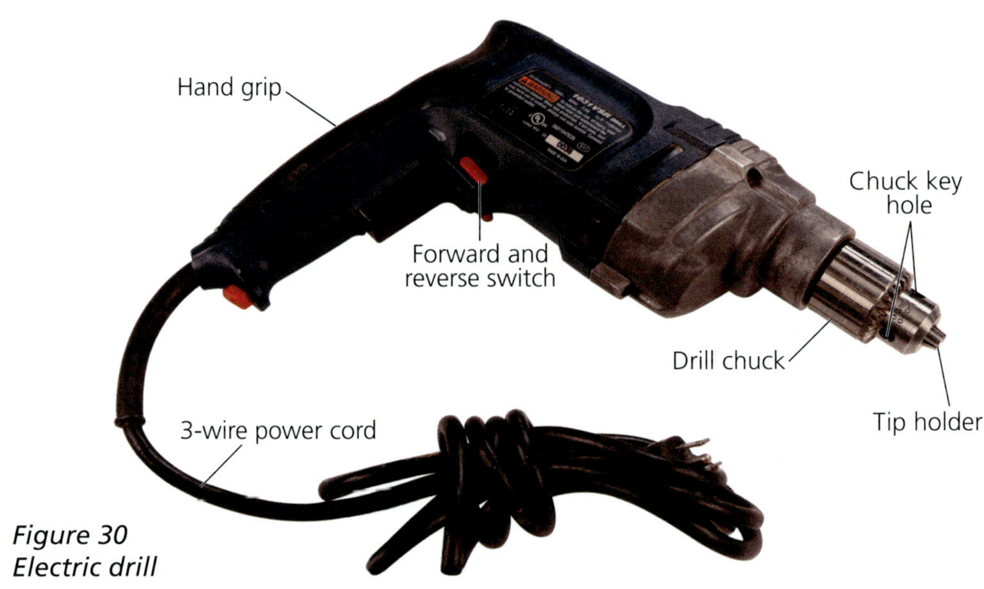

Figure 30
Electric drill

Because power drills are rotating tools, care must be taken to prevent hair, clothing, or dangling jewelry from being caught in the tool. Larger drills have higher torque and must be held securely to prevent injury.

Because power drills are rotating tools, care must be taken...

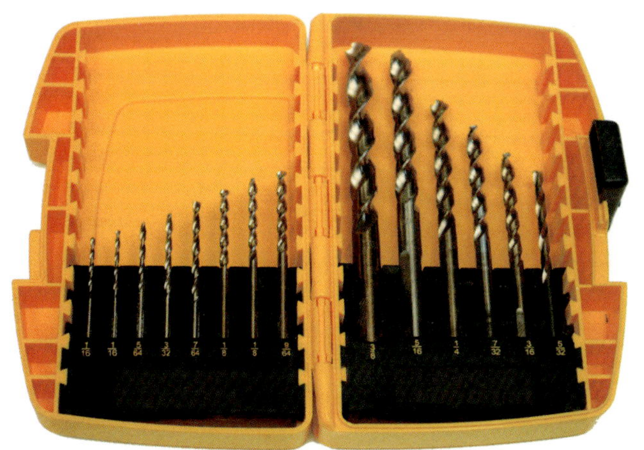

Figure 31
Drill index

Drill Index A drill index, like the one shown in Figure 31, is a container for drill bits. The bits are arranged and marked according to size. This makes it easier to select the exact size of bit you need for a particular task.

Figure 32
Twist bits

Twist Bit A twist bit, like those shown in Figure 32, has flutes that spiral around the shank to carry away waste material that has been cut from the bottom of the bore. The twist bit has a pointed tip that breaks the surface of the material when the bit is rotated at high speeds. Usually, twist bits are used for drilling small holes in either metal or wood.

Spade Bit A spade bit, like those shown in Figure 33, is shaped a little like an ordinary garden spade. It has a round shank that broadens into a flat head. At the tip of the head is a long brad point with two cutting surfaces. The spade bit is used for drilling rough holes ranging from 1/4″ to 1 1/2″ in diameter. The holes bored by a spade bit tend to be a little rougher and less accurate than those produced by a twist bit.

Figure 33
Spade bits

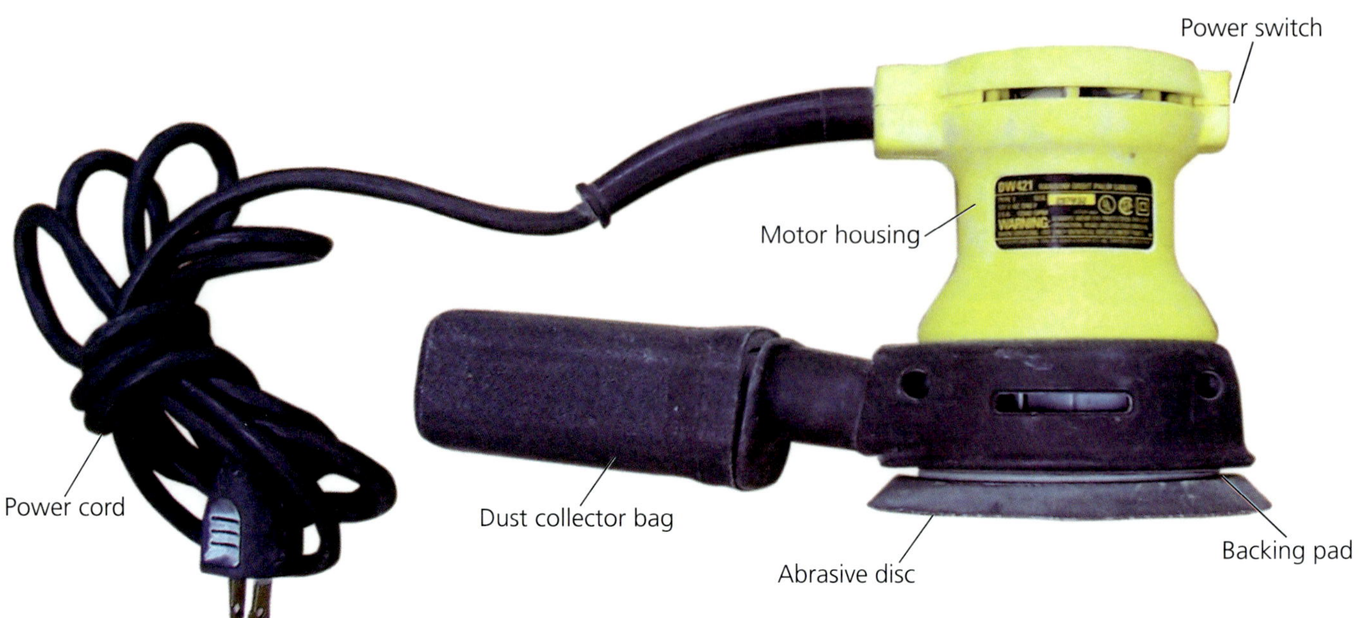

Figure 34
Random orbital sander

Random Orbital Sander

Shown in Figure 34, a random orbital sander is a handheld power tool that moves an abrasive disc in a motion that has no specific pattern. The random pattern of the sanding motion leaves fewer cross-grain scratches, which makes this type of sander useful for fine sanding before stains and finishes are applied to a surface. Random orbital sanders are equipped with round pads that hold abrasive discs in place. Ranging in size from 4″ to 8″ in diameter, these discs are disposable and can be quickly replaced when worn or when a different grit of abrasive disc is needed.

Pneumatic Staple Gun

Unlike ordinary staplers, which are hand-powered tools, pneumatic staple guns rely on high-pressure compressed air to drive staples into a material. An example of a pneumatic staple gun is shown in Figure 35. The compressed air is supplied to the staple gun through a hose from an electric compressor or pump to the pneumatic coupling. A staple gun is designed with a contact element, also called a foot, which when pressed against the surface of a material, disengages a safety catch and allows the staple gun to insert a staple when the trigger is pulled. This makes the process of inserting staples quick and easy. The staples are held in a long metal bar called a magazine. However, since pneumatic staple guns fire staples with considerable force, they can cause serious injury if used improperly. Malfunctioning staple guns can be very dangerous and must be taken out of service immediately.

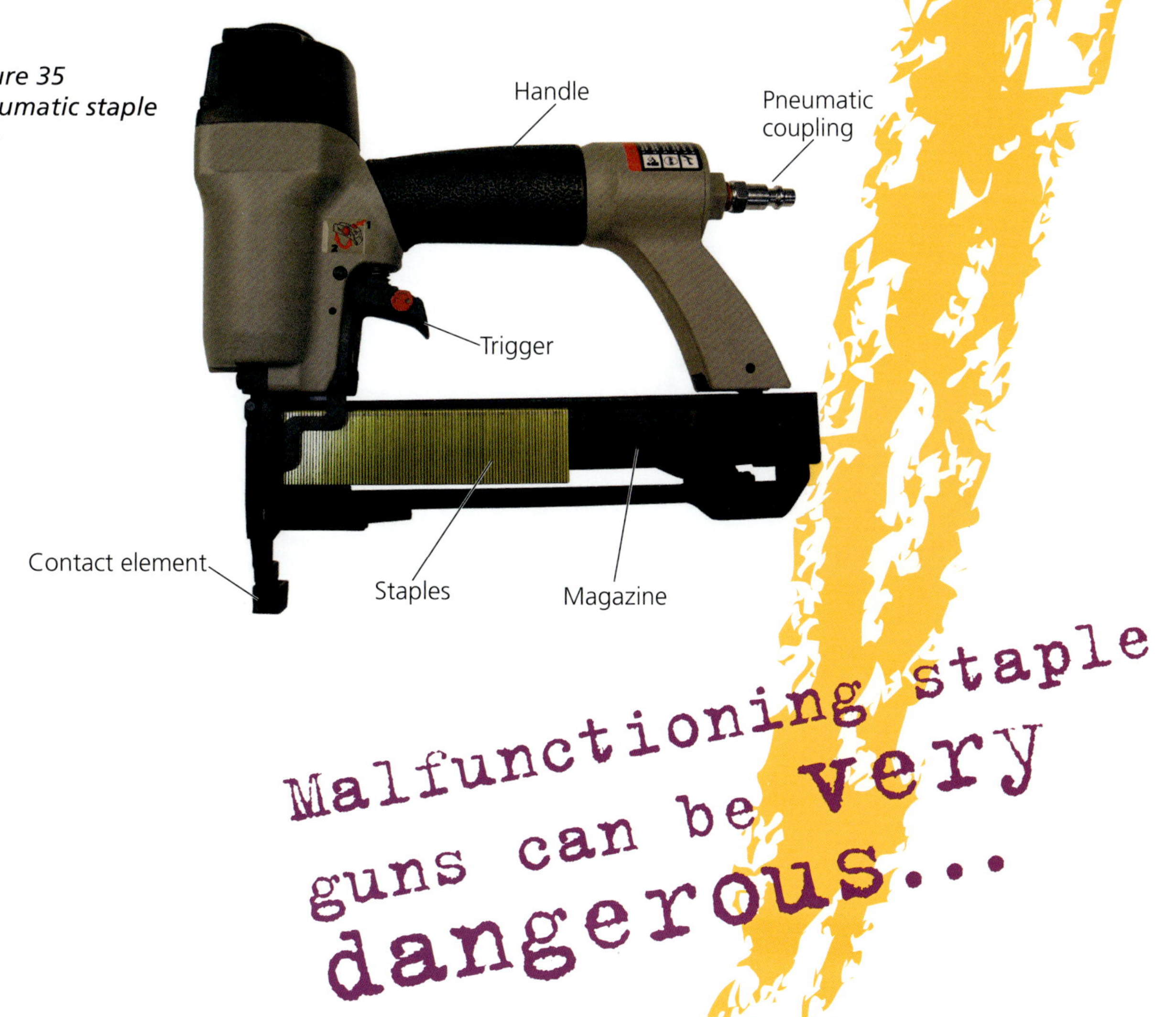

Figure 35 Pneumatic staple gun

Pneumatic Nail Gun

Pneumatic nail guns work much like pneumatic staplers. However, instead of staples, pneumatic nail guns use compressed air to drive nails and brads. When the nose, or contact element, is pressed against the material and the trigger is pulled, a nail or brad is inserted. The insertion pressure of a nail gun must be carefully adjusted to prevent damaging the surface. Like other power tools, nail guns are dangerous when used improperly. Nail guns are designed for a particular type of nail. Loading the gun with the wrong type of nail may result in a malfunction. Malfunctioning nail guns should be taken out of service immediately.

Figure 36
(A) Pneumatic framing nail gun

There are two types of nail guns. Framing nail guns, shown in Figure 36A, are used primarily for framing. Finishing nail guns, shown in Figure 36B, are used for assembling finished products. Framing nail guns typically use heavier gauge nails while finishing nail guns use less noticeable smaller gauge nails. Both nail guns are connected to the pneumatic power supply through the pneumatic coupling.

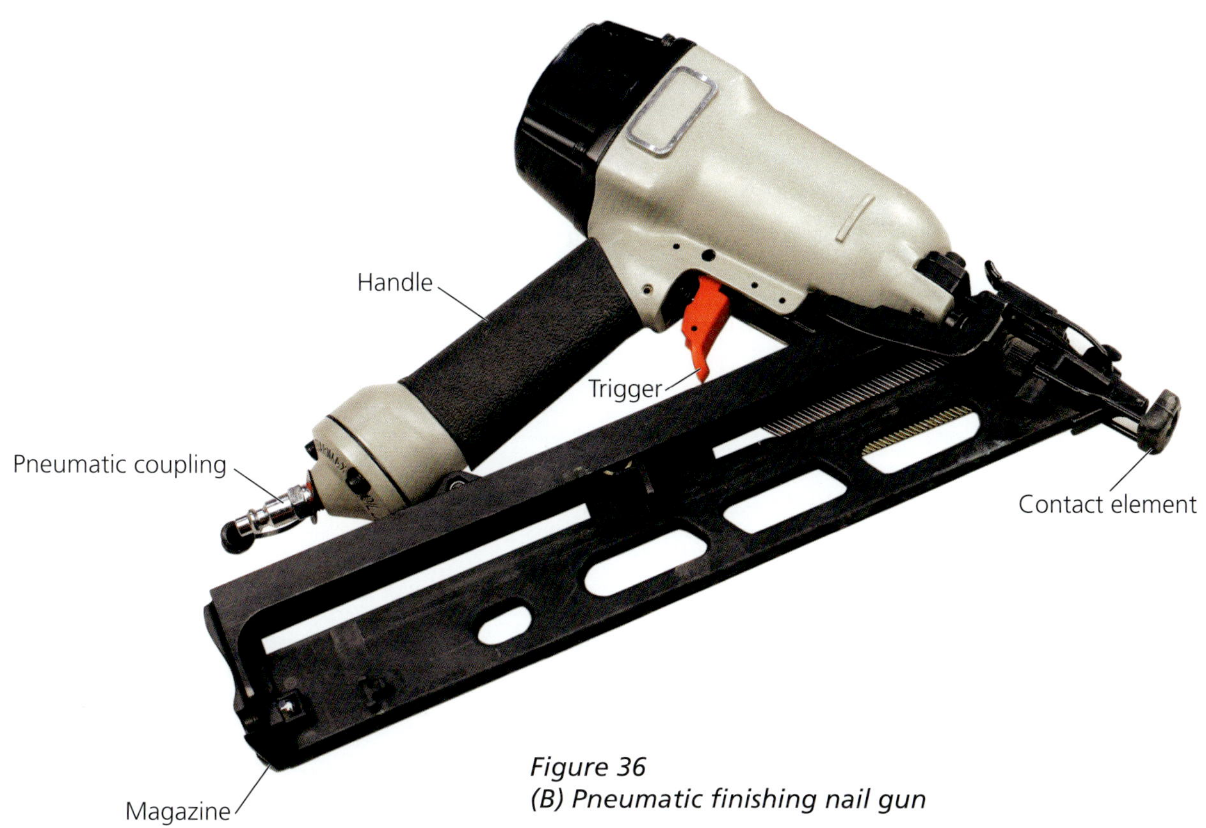

Figure 36
(B) Pneumatic finishing nail gun

...pneumatic nail guns use compressed air to drive nails and brads.

Safety rules that apply to other power tools also apply to routers...

Router

Used for milling the surface or edges of wood and other wood-like materials, a router is a cutting tool with a revolving vertical shaft, or spindle, and a cutter. An example of a router is shown in Figure 37. The cutter, or bit, fits into a collet and is rotated very rapidly by the router's electric motor. The spinning bit can bore or slice wood and other materials into almost any desired shape. Router bits are available in many different sizes, shapes, and types, and each one is designed to make a different type of cut. The bit selected depends on the sort of work that must be done.

Routers are difficult to operate and can be dangerous. Safety rules that apply to other power tools also apply to routers. When working with a router, it is especially important to select the correct bit for the job, to keep hands and fingers clear of the rotating bit, and to keep a firm hold on the router while it is in use.

Figure 37
Router

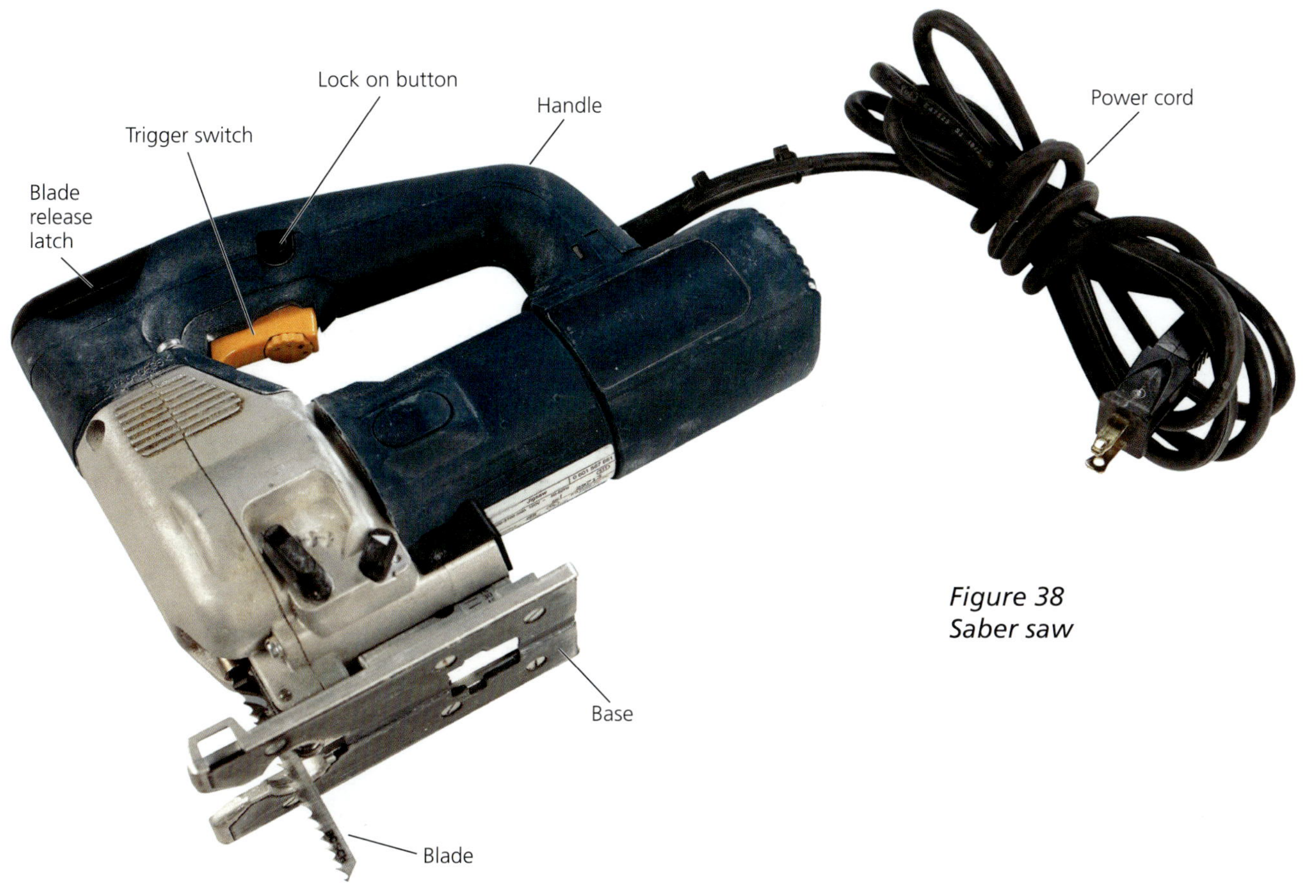

Figure 38
Saber saw

Saber Saw

A saber saw, also called a jigsaw, is a handheld power cutting tool with an elongated blade that gives it a sword-like appearance. Saber saws, like the one shown in Figure 38, are often used for cutting irregular shapes such as curves and circles. The blade of a saber saw moves up and down or in an oval motion as it slices into the material. Different blades are used for cutting wood, metal, and plastic. Many saber saws offer variable speed settings and are equipped with a trigger lock for continuous operation. Like any power tool, saber saws can be dangerous, and you must use caution when working with one. Many of the same safety rules that apply to circular saws and sliding compound miter saws also apply to saber saws.

Screw Gun

A screw gun is a power screwdriver used to drive or remove screws in wood, metal, or drywall. An example is shown in Figure 39. The screw gun can be set to rotate either clockwise or counterclockwise so the gun can be used to both insert screws and remove them. Screw guns can operate at variable speeds and are equipped with interchangeable tips that will fit most screw heads. Some screw guns have an adjustable nose cone for driving screws to a preset depth.

Figure 39
Screw gun

4 Materials

Carpenters work with a wide variety of materials. In *Career Connections: Project Book 1*, you were introduced to a number of materials frequently used by professional carpenters. You will be using some or all of these materials again while completing the projects presented later in this book. These materials are listed below along with a brief review of each.

Figure 40
Cedar

Cedar

Since it is highly resistant to decay, cedar is often selected for products that will likely be exposed to weather. The durability of cedar makes it an excellent choice for use in outdoor areas. Cedar can be easily cut, but it often splits and doesn't hold fasteners well. Figure 40 shows an example of cedar.

Douglas Fir

Douglas fir tends to be a strong, straight, and even-grained wood with a medium to coarse texture. Ranging in color from white to yellow brown, it is very attractive and useful for many different purposes. Douglas fir machines well and can be easily nailed, screwed, or glued. It also receives preservatives, stains, paints, and varnishes well. Unlike cedar, Douglas fir has low resistance to decay. Figure 41 shows an example of Douglas Fir.

Figure 41
Douglas fir

Oak

Known for its straight grain and attractive brown color, oak can be easily worked with machine tools and holds fasteners and adhesives well. However, oak is difficult to work by hand, and it has a tendency to splinter and chip when cut against the grain. Oak is often so hard and dense that the holes for nails and screws must be predrilled. An example of oak is shown in Figure 42.

Figure 42
Oak

Figure 43
Pine

Pine

Because it is plentiful and relatively inexpensive, pine lumber ranks among the materials most often used in carpentry. Carpenters make use of several species of pine. These species vary somewhat, but most offer a straight, fine grain and good working characteristics. Pine is cut and machined easily and receives nails, screws, paints, and other finishes fairly well. The chief disadvantage of pine is that it is not as strong or durable as some other woods. An example of pine is shown in Figure 43.

Figure 44
Plywood

Figure 45
Oriented strand board

Plywood

Plywood, shown in Figure 44, ranks among the most frequently used engineered materials. Manufactured from thin layers of wood that are glued together at right angles using pressure, heat, and bonding agents, plywood is stronger than natural wood. To improve its appearance, an attractive veneer is often used for the top layer of the plywood. Plywood is available in a variety of sizes, but is commonly sold in 4′ x 8′ sheets. Plywood is available in thicknesses from 1/4″ to 1 1/4″.

Oriented Strand Board

Oriented strand board (OSB), as shown in Figure 45, is an engineered wood panel made of layers of thin wood shavings that are glued together at right angles using pressure, heat, and bonding agents. The shavings or strands are placed or "oriented" in such a way as to increase the strength of the board. Relatively inexpensive, OSB is usually considered an acceptable substitute for plywood. OSB is available in thicknesses ranging from 1/4″ to 3/4″.

Pressboard

Pressboard, also called hardboard, is a dense engineered panel made of small wood fibers that are joined under heat and pressure. Shown in Figure 46, pressboard is often used for cabinet backs and drawer bottoms or wherever the finished appearance of the component is not important. Pressboard is available in a variety of sizes, but is commonly sold in 4′ x 8′ sheets, 1/8″ and 1/4″ thick.

Figure 46
Pressboard

5 Fasteners

Regardless of the materials used, the components of a carpentry project must be held together with fasteners of some sort. These may include metal fasteners such as nails and screws or adhesives such as yellow wood glue. You used a variety of fasteners and adhesives to complete the projects outlined in *Career Connections: Project Book 1*, and you will be using them again while working on the projects in this book. These fasteners and adhesives are listed below along with a brief review of each.

Nails

Nails are thin, straight metal shafts with a point at one end and a head at the other. They are driven into materials using hammers or nail guns. As a nail is being driven, the point pushes aside wood fibers which then close around the shaft and grip it firmly. Nails come in a variety of sizes ranging from small 2d, or "two penny," nails to large 20d, or "twenty penny," nails. Most nails have sharp points although some have rounded or blunt points. Nails with sharp points tend to hold better while those with blunt points are less likely to split the wood while being driven. Examples of nails can be seen in Figure 47.

Figure 47
Nails

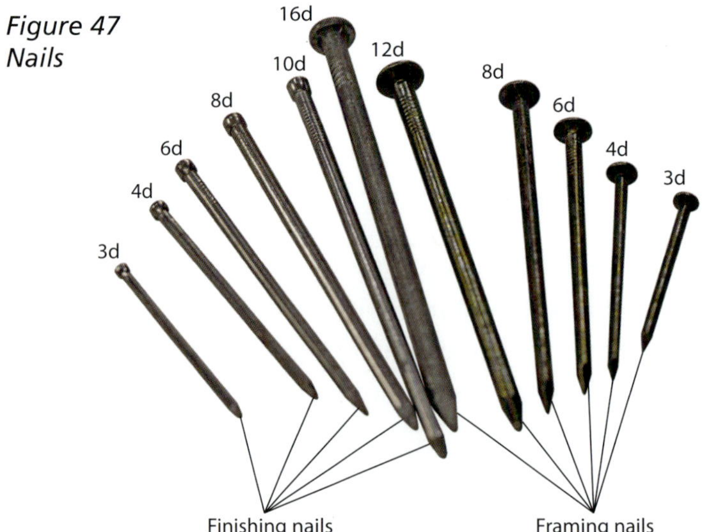

Nails are manufactured in a variety of lengths and with many different kinds and shapes of heads. Framing nails have a flat head and are designed to cover the nail hole. Finishing nails have very small, slightly concave heads designed to be driven beneath the surface of the wood with a nail set.

Brads

As shown in Figure 48, a brad is a thin gauge nail used to avoid splitting the grain of wood. Typically used only for light assembly work, brads look much like small finishing nails. They should not be used for structural components, that is, for parts that are critical to the stability or strength of the product.

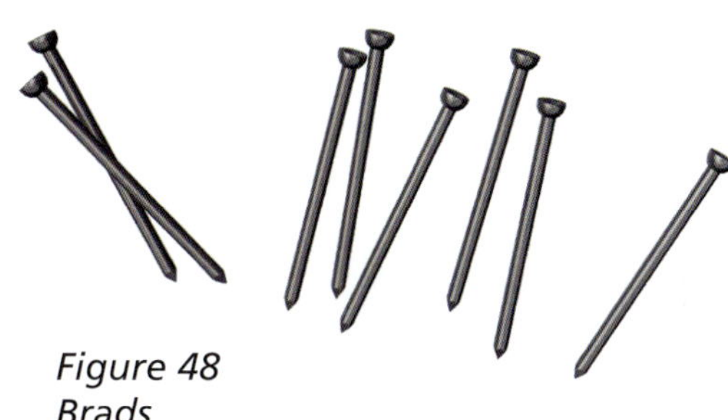

Figure 48
Brads

Screws

As shown in Figure 49, screws have spiral threads that bite into wood or other materials. This gives them greater holding power than nails. Instead of being hammered into place, screws are twisted into place with a manual screwdriver or a screw gun. As with nails, the holding power of a screw depends in part on its size. The top portion of the screw is called the head. Screw heads are designed to accept one or another of several different types of screwdriver tips. The screwdriver tip must fit the screw head snugly. Carpenters often use wood screws which are threaded only along the lower portion of the shaft. They also sometimes use construction screws which have a sloped or bugle shaped head that helps the screw head fit smoothly into the material.

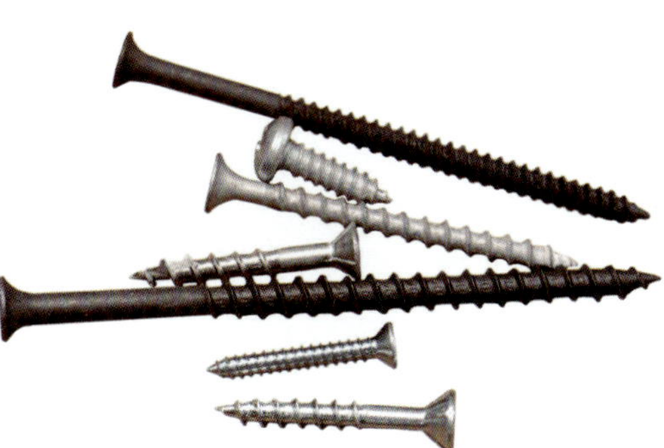

Figure 49
Screws

...components of a carpentry project must be held together with fasteners of some sort.

Yellow Wood Glue

Yellow wood glue bonds mechanically, flowing into the pores of the parts it is holding together. Yellow wood glue is shown in Figure 50. This makes it ideal for use on wood and similar materials that have abundant pores. When using yellow wood glue, it is important to avoid a buildup of excess adhesive around the joint. After the glue is spread on the wood surface the pieces are clamped together until the glue dries.

Figure 50
Yellow wood glue

Chapter Check

On a separate piece of paper, write your answers to the following questions:

Multiple Choice

1. Which of the following hand tools has a beveled end?
   a) nail set
   b) claw hammer
   c) compass
   d) chisel

2. Which of the following hand tools is used to mark long straight lines?
   a) chalk box
   b) straight edge
   c) tape measure
   d) combination square

3. Which of the following power tools is specially designed for cutting at an angle?
   a) saber saw
   b) router
   c) miter saw
   d) circular saw

4. Which of the following power tools has a revolving vertical shaft and a cutter?
   a) saber saw
   b) router
   c) miter saw
   d) circular saw

5. Which of the following materials is an excellent choice for outdoor use?
   a) Douglas fir
   b) pine
   c) oak
   d) cedar

Chapter Check

Fill in the Blank

6. A(n) ______________________ is an ideal hand tool for rough shaping wood.
7. A(n) ______________________ bit is used for drilling holes from ¼″ to 1 ½″ in diameter.
8. Holes for nails and screws must be predrilled in ______________________ because it is such a hard and dense wood.
9. A(n) ______________________ is a fastener used to avoid splitting the grain of wood when doing light assembly work.
10. A(n) ______________________ nail is designed with a flat head to cover the nail hole.

True or False

11. An accidental startup of a power tool can occur if you attempt to change the blade without turning off the power. *T or F*
12. Plywood can be over an inch thick. *T or F*
13. A pneumatic nail gun can accommodate any type of nail. *T or F*
14. The material being cut by a circular saw should be well supported so that it will not bind the blade. *T or F*
15. A nail's point pushes aside wood fiber as it is driven into the material, which then closes around the shaft of the nail and grips it to hold it in place. *T or F*

CHAPTER 3

SAWHORSE

CONTENTS

Introduction

Sawhorses **assist** carpenters by **supporting** pieces of **lumber...**

A sawhorse is a very practical item that is used often by professional carpenters. Sawhorses assist carpenters by supporting pieces of lumber and other components while they are measured, laid out, cut, and fastened. Usually consisting of four braced legs and a sturdy top, a sawhorse has a relatively simple design. However, you may find that building one is more complicated than you might think. For instance, the legs must be measured, cut, and assembled very carefully. Otherwise the legs will not hold the top of the sawhorse straight and level and the sawhorse will wobble. Typically, sawhorses built for use with power tools will be made taller than those for use with hand saws. Shorter legs allow for more leverage when using a hand saw. The height of the sawhorse is also determined by the height of the carpenter who will use it.

This chapter includes detailed instructions for building three different types of sawhorses. The first sawhorse has notches, called gains, which hold the legs in place. The second sawhorse uses an I-beam in its construction. The third sawhorse is made with a beveled edge top. Each of these projects will provide you with valuable experience in standard carpentry processes such as print reading, measuring, layout, cutting, and assembly. You will also get practice safely handling and using common carpentry hand tools, power tools, materials, and fasteners.

What's New?

The projects in this chapter are intended to expand your growing base of carpentry knowledge and skills. Completing a notched-top sawhorse, I-beam sawhorse, or beveled-top sawhorse will require you to learn about several new processes and tools. Among these are the following:

Figure 1
Stair gauge clamps

Stair Gauge Clamps Typically used in pairs, stair gauges are small clamps that can be tightened and locked into position with a thumbscrew. Stair gauges are often used in combination with a framing square to make it easier to repeat precise measuring and layout operations. Figure 1 shows an example of stair gauge clamps.

Compound Angle Cut As its name suggests, a compound angle is a combination of two angles. In carpentry, a compound angle cut is angled in two separate planes. Since compound angle cuts are more complex than other cuts, they must be carefully measured and then cut accurately.

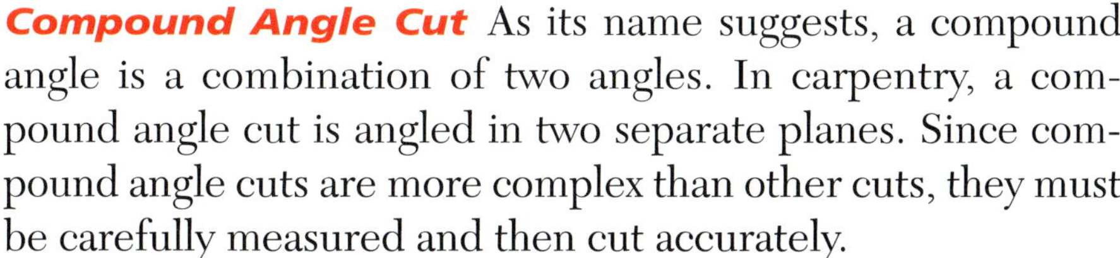

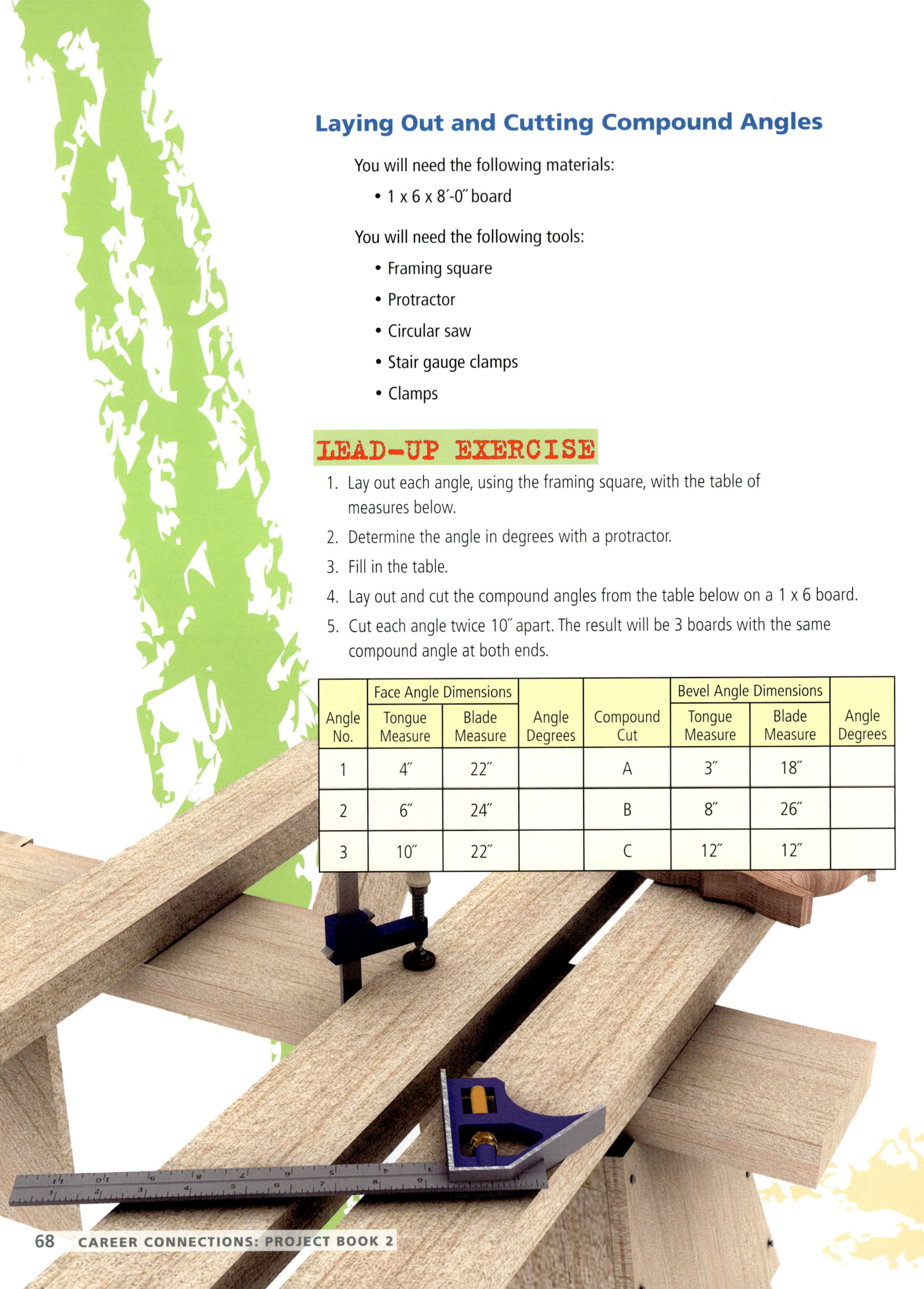

Laying Out and Cutting Compound Angles

You will need the following materials:

- 1 x 6 x 8′-0″ board

You will need the following tools:

- Framing square
- Protractor
- Circular saw
- Stair gauge clamps
- Clamps

LEAD-UP EXERCISE

1. Lay out each angle, using the framing square, with the table of measures below.
2. Determine the angle in degrees with a protractor.
3. Fill in the table.
4. Lay out and cut the compound angles from the table below on a 1 x 6 board.
5. Cut each angle twice 10″ apart. The result will be 3 boards with the same compound angle at both ends.

Angle No.	Face Angle Dimensions		Angle Degrees	Compound Cut	Bevel Angle Dimensions		Angle Degrees
	Tongue Measure	Blade Measure			Tongue Measure	Blade Measure	
1	4″	22″		A	3″	18″	
2	6″	24″		B	8″	26″	
3	10″	22″		C	12″	12″	

1 Building a Notched-Top Sawhorse

To construct a notched-top sawhorse you will use a variety of measuring, cutting, and fastening tools to cut and assemble components that include both dimensional lumber and plywood. The construction of the sawhorse will be accomplished in three stages or procedures. First you will lay out and cut the legs. Next you will lay out and cut the top, and finally, you will attach the legs, gussets, and side spreaders.

You will need the following materials:

- (1) 1 x 4 x 8′-0″ lumber
- (1) 2 x 6 x 4′-0″ lumber
- (1) 1 x 4 x 12′-0″ lumber
- (1) ½″ x 12″ x 24″ plywood
- (20) 1 ⅝″ construction screws
- (16) 1 ¼″ construction screws

The construction of the sawhorse will be accomplished in three stages...

You will need the following tools:

- Circular saw
- Clamps
- Combination square
- Drill index
- Electric drill
- Extension cord
- Framing square
- Hammer
- Hand saw
- Pencil
- Tape measure
- Screw gun
- Sliding T-bevel
- Wood chisel

Notched Sawhorse

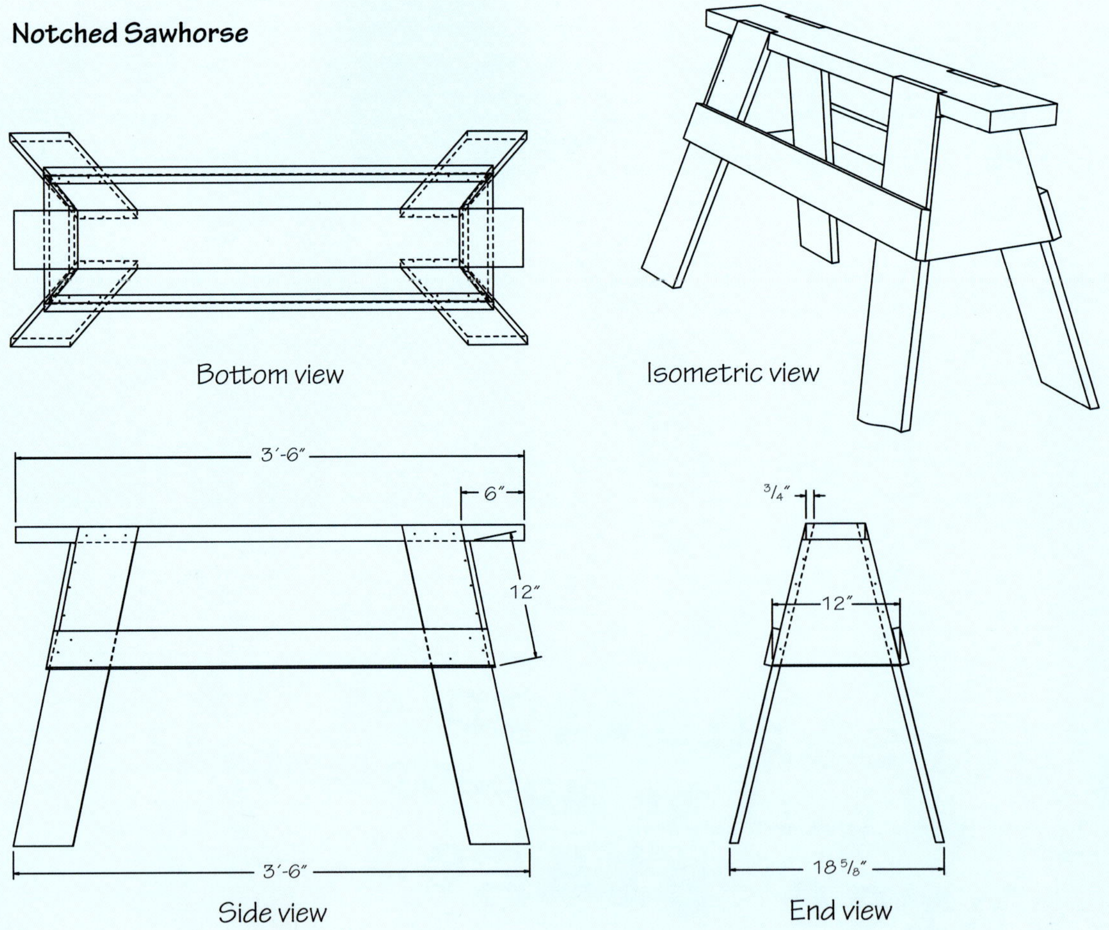

Bottom view

Isometric view

3′-6″

6″

12″

3′-6″

Side view

3/4″

12″

18 5/8″

End view

Lay Out and Cut the Legs

The legs of the sawhorse must be measured and cut precisely. If they are not of equal length, the sawhorse may wobble when placed on a level surface. You will use a framing square, tape measure, and sliding T-bevel to measure the legs and then cut them with a circular saw.

Figure 2
Figure 2 illustrates step 2

PROCEDURE

1. Place the 1 x 4 x 12′-0″ on the supported work surface.
2. Position the framing square approximately 2″ away from one end of the 1 x 4, aligning the 2 7⁄16″ mark on the tongue and the 12″ mark on the body with the edge of the 1 x 4. See Figure 2.
3. Draw a line along the tongue across the 1 x 4. See Figure 3.
4. Hold the handle of the sliding T-bevel against the edge of the 1 x 4 and align the blade with the line drawn in step 3. Tighten the locking device on the sliding T-bevel. See Figure 4.
5. Set the base plate of the circular saw for a 12° bevel.

Figure 3
Figure 3 illustrates step 3

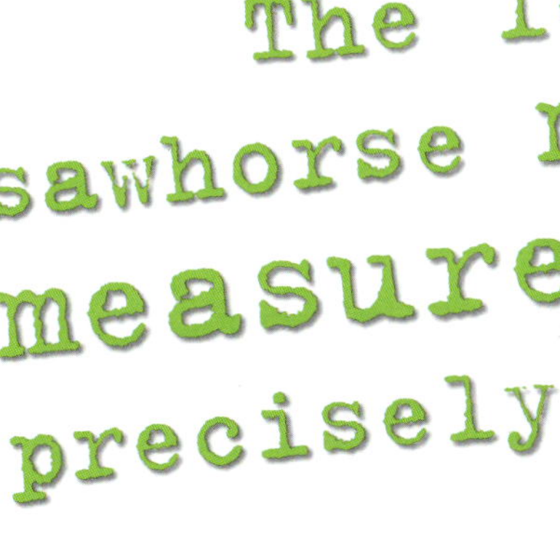

Figure 4
Figure 4 illustrates step 4

Figure 5
Figure 5 illustrates step 6

6. Cut along the line drawn in step 3. See Figure 5. This is a compound angle cut.
7. Hook the end of the tape measure on the long point of the cut made in step 6 and measure along the edge of the 1 x 4 a distance of 2′ 7 ¼″ and place a mark. See Figure 6.
8. Draw a line across the 1 x 4 parallel to the cut made in step 6, using the sliding T-bevel. See Figure 7.

Figure 6
Figure 6 illustrates step 7

Figure 7
Figure 7 illustrates step 8

9. Cut along the line made in step 8 using the circular saw. See Figure 8. Be certain to start the cut from the same edge of the board as the cut made in step 6.
10. Repeat steps 7–9 three times to cut the remaining three legs.
11. Mark the legs 1, 2, 3, and 4. See Figure 9.

Figure 8
Figure 8 illustrates step 9

Figure 9
Figure 9 illustrates step 11

Lay Out and Cut the Top

The second stage of the notched-top sawhorse project involves laying out and cutting the top. First you will carefully measure, mark, and cut the top board. Then you will measure and cut the notches, or gains, that will help hold the legs in place. See Figure 10 for an example of the notched-top in a notched-top sawhorse. Finally, you will remove waste material and smooth the notches using a chisel.

Figure 10
Notched-top sawhorse

Figure 11
Figure 11 illustrates step 1

PROCEDURE

1. Measure and mark the top board 3′-6″. See Figure 11. Then cut the board at that mark.
2. Measure back 6″ from each end and square a line across the width of the board. See Figure 12.
3. Set the sliding T-bevel at the angle used to lay out the legs and mark a line on each side of the board at the marks made in step 2. See Figure 13.

Figure 12
Figure 12 illustrates step 2

Figure 13
Figure 13 illustrates step 3

4. Align the edge of the leg cut earlier with a line made in step 3 and draw a fine line on the other edge of the leg. Mark this gain with the number on the leg used to trace the gain. See Figure 14.
5. Repeat step 4 for the remaining three legs.
6. Draw a square line across the top face of the top board from the line made on the edge of the board in step 4. See Figure 15. Repeat for all legs.
7. Set the combination square at ¾″ and scribe a line parallel to each edge of the top board between the square lines on the top side of the top board. See Figure 16. Scribe these lines on both ends of the top board.
8. Turn the top board over on the work surface.

Figure 14
Figure 14 illustrates step 4

Figure 15
Figure 15 illustrates step 6

Figure 16
Figure 16 illustrates step 7

Figure 17
Figure 17 illustrates step 9

9. Draw square lines across the bottom face of the top board from the lines made on the edge of the board in steps 3 and 4. See Figure 17. Repeat for all legs.
10. Set the combination square to $\frac{7}{16}$″ and scribe a line parallel to each edge on the bottom face of the top board between the square lines drawn in step 9. Mark these lines on both ends of the top board. See Figure 18.
11. Clamp the top board to the work surface on edge. See Figure 19.

Figure 18
Figure 18 illustrates step 10

Figure 19
Figure 19 illustrates step 11

Figure 20
Figure 20 illustrates step 12

12. Cut the gains using the hand saw, staying on the waste side of the angled lines on the edge of the board. Be careful to not cut beyond the lines laid out on the top and bottom of the top board. See Figure 20.
13. Make additional saw cuts between the cuts made in step 12 to create relief cuts. See Figure 21.

Figure 21
Figure 21 illustrates step 13

Use a wood chisel to remov the waste material.

Figure 22
Figure 22 illustrates step 14

14. Use a wood chisel to remove the waste material of the gain and to smooth the gain. See Figure 22.
15. Repeat steps 13 and 14 to complete the other gains. See Figure 23.

Figure 23
Completed gains

Attach the Legs, Gussets, and Side Spreaders

Once the legs and top of the sawhorse have been measured, cut, and prepared, you can begin the process of assembling the sawhorse. You will use an electric drill to drill pilot holes and a screw gun to fasten the components with construction screws. To support the legs and hold them at the proper angle you will also cut and install plywood end pieces, called gussets, and side spreaders.

PROCEDURE

1. Layout and drill three pilot holes parallel to the top end of each leg. Drill the holes ¾″ deep. See Figure 24.
2. Align two legs in the gains on one side of the top board. Be sure that the leg tops are flush with the top board. See Figure 25.
3. Attach the two legs to the top board with three 1 ⅝″ construction screws per leg. See Figure 26.
4. Turn the assembly over and repeat steps 2 and 3 for the remaining legs.
5. Cut two pieces of ½″ x 12″ x 12″ plywood. These plywood end pieces will eventually become the gussets.
6. Place one plywood end piece against the bottom of the top board at one end.

Figure 24
Figure 24 illustrates step 1

Figure 25
Figure 25 illustrates step 2

Figure 26
Figure 26 illustrates step 3

7. Mark the outside edge of the legs at the top of the plywood end piece. See Figure 27.
8. Align the 12″ mark on the body of the framing square and the 2 ⅝″ mark on the tongue of the framing square with the edge of a piece of scrap material and draw a line along the tongue. See Figure 28.
9. Set a sliding T-bevel to the line drawn in step 8. See Figure 29.
10. Transfer the angled lines to the plywood end piece, using the sliding T-bevel, from the top marks made in step 7. Use a framing square to extend the line to the bottom of the plywood end piece.
11. Cut along the angled lines drawn in step 10 with the circular saw.

Figure 27
Figure 27 illustrates step 7

Figure 28
Figure 28 illustrates step 8

Figure 29
Figure 29 illustrates step 9

Figure 30
Figure 30 illustrates step 12

12. Place the saw horse upright on a flat surface with the spread of the legs set at 18 ⅝″ at each end. See Figure 30.
13. Place the plywood end piece under the top board on one end, and adjust the legs to align the edges with the end piece and clamp the end piece to the legs. See Figure 31.
14. Fasten the end piece to the legs with eight 1 ¼″ construction screws.
15. Repeat steps 6–14 for the other end piece at the other end of the horse.
16. Cut the 1 x 4 x 8′-0″ in half with a circular saw.
17. Spread the legs to 3′-6″. See Figure 32.
18. Position one piece of the 1 x 4 cut in step 16 against the two side legs of the horse aligning the bottom of the 1 x 4 with the bottom of the plywood gussets.

Figure 31
Figure 31 illustrates step 13

Figure 32
Figure 32 illustrates step 17

Figure 33
Figure 33 illustrates step 19

19. Draw lines on the 1 x 4 side spreader along the edge of the gussets. See Figure 33.
20. Cut the side spreader to length, with the circular saw, along the lines drawn in step 17.
21. Attach the side spreader to the legs with four 1 ⅝″ construction screws. See Figure 34. Use two per leg.
22. Repeat steps 17–21 for the second side spreader.
23. Check the sawhorse for wobble on a flat surface and adjust leg length as needed.

Figure 34
Figure 34 illustrates step 21

alternate project 2 I-Beam Sawhorse

An I-beam sawhorse will provide strong and stable support for heavy materials that need to be cut or worked. This type of sawhorse draws its considerable strength from its top section which is shaped like the steel I-beams used to construct buildings and bridges. Building the I-beam sawhorse will require you to successfully complete five separate procedures. You will make the I-beam top, cut the legs, attach the legs to the I-beam, prepare and attach the gussets, and attach the top 2 x 4.

You will need the following materials:

- (1) 1 x 4 x 12′-0″ lumber
- (1) 1 x 4 x 8′-0″ lumber
- (1) 2 x 4 x 8′-0″ lumber
- (1) ½″ x 12″ x 12″ or thicker plywood
- (44) 1 ⅝″ construction screws

You will need the following tools:

- Circular saw
- Clamps
- Drill index
- Electric drill
- Extension cord
- Framing square
- Hammer
- Pencil
- Tape measure
- Screw gun
- Stair gauge clamps

An I-beam sawhorse will provide strong and stable support for heavy materials...

Make the I-Beam

Making the I-beam portion of the sawhorse is a fairly simple but very important procedure. The I-beam is made from 4′-0″ lengths of 1 x 4 and 2 x 4 lumber. The 1 x 4 pieces are arranged on the top and bottom of the 2 x 4 piece and then firmly attached using construction screws.

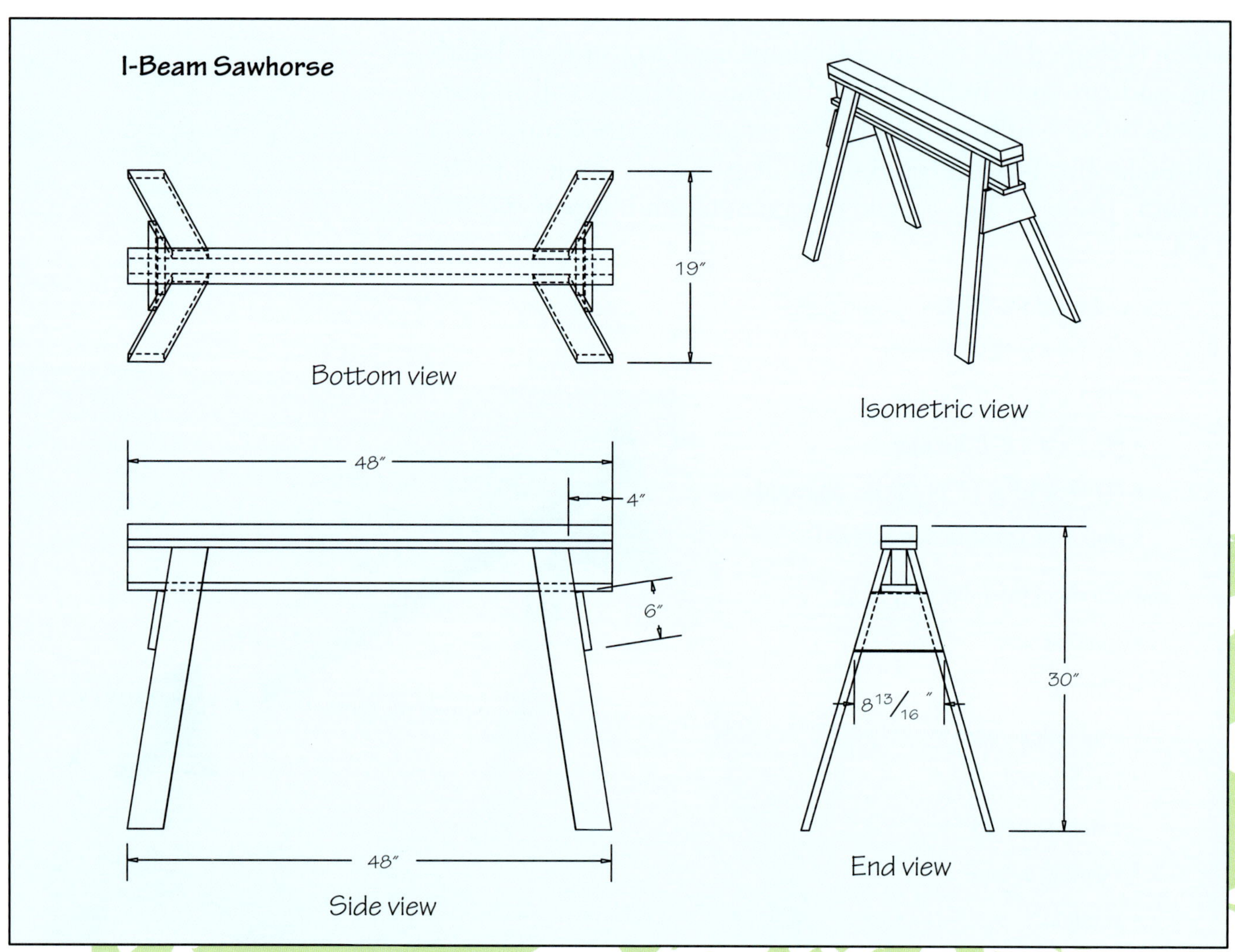

Figure 35
Figure 35 illustrates step 2

PROCEDURE

1. Cut the 1 x 4 x 8′-0″ and the 2 x 4 x 8′-0″ in half.
2. Center and attach each 1 x 4 from step 1 on the top and bottom of one piece of the 2 x 4 from step 1 with eight 1 ⅝″ construction screws. See Figure 35.

Cut the Legs

The sawhorse will stand on four legs, which must be of exactly the same length. To make sure the legs are identical, they are measured and marked using a framing square and stair gauges. The stair gauges make it relatively easy to repeat precise layouts any number of times. Once they are properly marked, the legs are cut with a circular saw set at an angle of 16°.

Figure 36
Figure 36 illustrates step 2

PROCEDURE

1. Place the framing square along the edge of the 1 x 4 x 12′-0″, aligning it at 1 ¾″ on the tongue and 12″ on the body.
2. Slide the stair gauges onto the framing square and clamp the gauges to the square, keeping the square on the aligned marks in step 1. See Figure 36.
3. Draw a line along the tongue of the framing square at one end of the 1 x 4 x 12′-0″. See Figure 37.
4. Set the base plate on the circular saw to an angle of 16°.
5. Cut along the line laid out in step 3 with the circular saw. See Figure 38.

Figure 37
Figure 37 illustrates step 3

Figure 38
Figure 38 illustrates step 5

6. Measure along the edge of the 1 x 4 a distance of 29″ from the long point of the cut made in step 5. See Figure 39.
7. Draw a line parallel to the cut made in step 5, using the framing square.
8. Cut along the line drawn in step 7, with the circular saw still set at 16°. Be certain that the beveled cut is parallel to the beveled cut made in step 5. See Figure 40.
9. Repeat steps 6, 7 and 8 to cut the remaining 3 legs for the sawhorse.

Figure 39
Figure 39 illustrates step 6

Figure 40
Figure 40 illustrates step 8

Attach the Legs to the I-Beam Top

Assembly of the sawhorse begins by attaching the legs to the I-beam top. As part of this process you will use an electric drill and twist bit to make pilot holes. The pilot holes will make it easier to insert the construction screws that will attach the legs firmly to the I-beam.

Figure 41
Figure 41 illustrates step 1

PROCEDURE

1. Measure 4″ from the end of the 2 x 4 I-beam and place a mark where the 1 x 4 meets the 2 x 4. See Figure 41.
2. Repeat step 1 for the opposite end and the other side of the I-beam top.
3. Drill three pilot holes parallel to the top end of each leg. See Figure 42.
4. Lay the I-beam assembly on its side.
5. Align the outside edge of one leg with a 4″ mark drawn earlier and the beveled cut of the leg firmly against the 1 x 4.

Figure 42
Figure 42 illustrates step 3

Figure 43
Figure 43 illustrates step 6

6. Attach the leg to the I-beam top with three 1 ⅝″ construction screws. See Figure 43.
7. Drill 2 pilot holes through the leg into the center of the lower 1 x 4.
8. Fasten with two 1 ⅝″ construction screws. See Figure 44.
9. Repeat steps 5–8 for the remaining three legs.

The pilot holes will make it easier to insert the construction screws...

Figure 44
Figure 44 illustrates step 8

Figure 45
Figure 45 illustrates step 1

Prepare and Attach the Gussets

The legs of the sawhorse are reinforced with braces known as gussets. The gussets are cut from plywood using a circular saw. Then the gussets are attached to the legs of the sawhorse with construction screws.

PROCEDURE

1. Stand the sawhorse upright on a flat surface. See Figure 45.
2. Check to make certain that all four legs contact the floor.
3. Cut the 12″ x 12″ plywood in half creating two 6″ x 12″ pieces. See Figure 46.
4. Place one 6″ x 12″ piece of plywood against the outside face of the legs and firmly against the bottom of the I-beam.

Cut the plywood in half
creating two pieces...

Figure 46
Figure 46 illustrates step 3 completed

5. Trace the outside edges of the legs, with a pencil, on the backside of the plywood. See Figure 47.
6. Cut the lines drawn in step 5 with the circular saw, creating one gusset.
7. Drill three pilot holes through the gusset into the center of each leg.
8. Attach the gusset to the legs with six $1\frac{5}{8}$″ construction screws. See Figure 48.
9. Repeat steps 4–8 for the other pair of legs.

Figure 47
Figure 47 illustrates step 5

Figure 48
Figure 48 illustrates step 8

Figure 49
Figure 49 illustrates step 2

Attach the Top 2 x 4

The top of the sawhorse consists of a single piece of 2 x 4 lumber that rests on top of the I-beam. The top 2 x 4 is attached to the I-beam using construction screws inserted from beneath the upper 1 x 4 I-beam component. Once the top 2 x 4 is in place and attached, the I-beam sawhorse is complete.

PROCEDURE

1. Lay the 2 x 4 on the supported work surface.
2. Put the sawhorse on the 2 x 4, aligning the ends and edges with the 1 x 4. See Figure 49.
3. Join the ends and edges with four 1 ⅝″ construction screws. Use two per side from the underside of the 1 x 4 of the I-beam. See Figure 50.

Figure 50
Figure 50 illustrates step 3

alternate project 3 Beveled-Top Sawhorse

Some sawhorses are made with a beveled top. Making this type of sawhorse requires that you use a circular saw to make long bevel cuts on the edges of the top. This will require an accurate layout and a steady hand while cutting. For extra support, the beveled-top sawhorse will also be reinforced with gussets and side spreaders. Building the beveled-top sawhorse will require you to successfully complete three separate procedures. You will lay out and cut the legs, lay out and cut the top, and attach the legs, gussets, and side spreaders.

You will need the following materials:

- (1) 2 x 6 x 4′-0″ lumber
- (1) 1 x 6 x 12′-0″ lumber
- (1) 1 x 4 x 8′-0″ lumber
- (1) ½″ x 12″ x 30″plywood
- (24) 1 ⅝″ construction screws
- (16) 1 ¼″ construction screws

This will require an accurate layout and a steady hand...

Beveled-Top Sawhorse

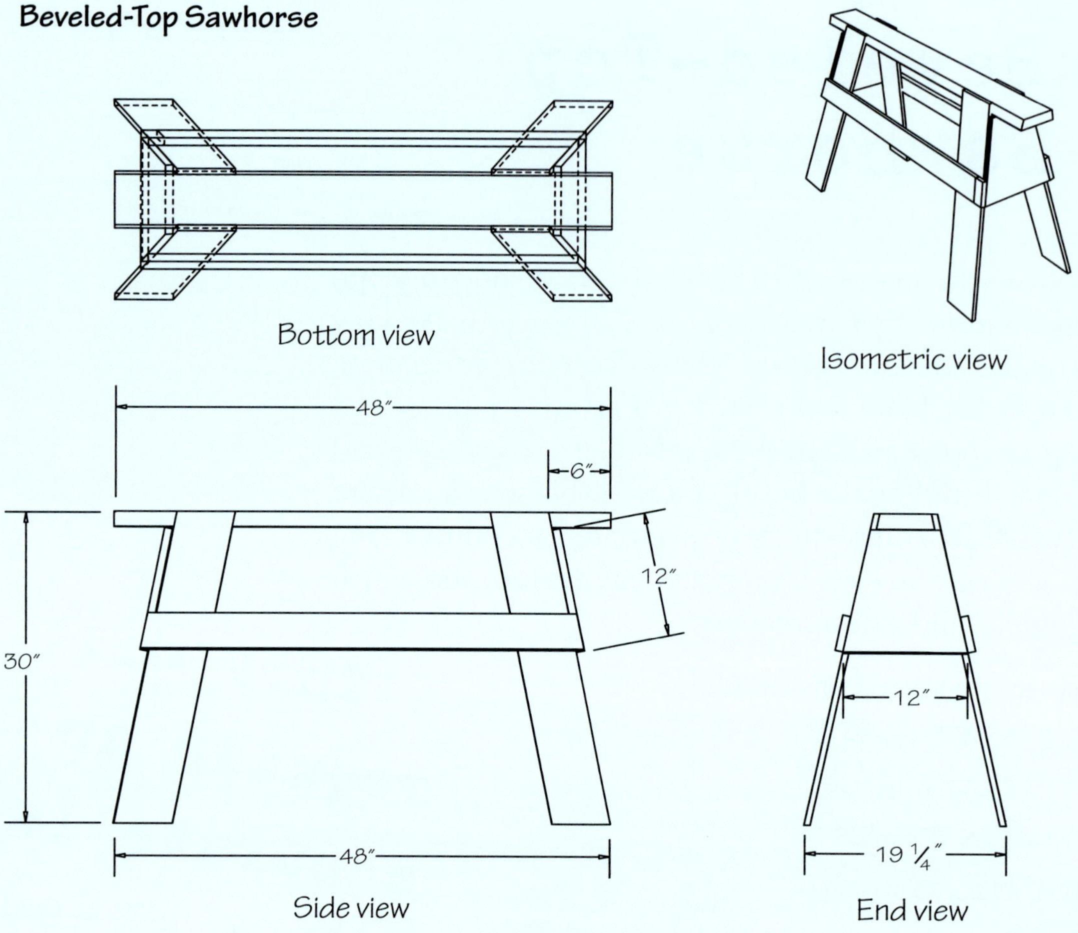

Bottom view

Isometric view

48″

6″

12″

30″

48″

Side view

12″

19 1/4″

End view

You will need the following tools:

- Circular saw
- Drill index
- Electric drill
- Extension cord
- Framing square
- Hammer
- Pencil
- Tape measure
- Screw gun
- Sliding T-bevel
- Speed square
- Straightedge
- Clamps

Lay Out and Cut the Legs

The legs of the sawhorse must be carefully measured and cut. Otherwise, the completed sawhorse will not stand straight and may wobble. You will use a pencil, tape measure, and sliding T-bevel to measure and mark the lumber. Then you will cut the legs using a circular saw.

PROCEDURE

1. Place the 1 x 6 x 12′-0″ on the supported work surface.
2. Position the framing square approximately 2″ away from one end of the 1 x 6. Align the 2 7⁄16″ mark on the tongue and the 12″ mark on the body with the edge of the 1 x 6.
3. Draw a line along the tongue across the 1 x 6. See Figure 51.
4. Hold the handle of the sliding T-bevel against the edge of the 1 x 6 and align the blade with the line drawn in step 3. Tighten the locking device on the sliding T-bevel.
5. Set the base plate of the circular saw for a bevel of 12°.
6. Cut along the line drawn in step 3.
7. Hook the end of the tape measure on the long point of the cut made in step 6 and measure along the edge of the 1 x 6 a distance of 2′-7 1⁄4″ and place a mark. See Figure 52.
8. Draw a line across the 1 x 6 parallel to the cut made in step 6, using the sliding T-bevel. See Figure 53.
9. Cut along the line made in step 8, using the circular saw. Be certain to start the cut from the same edge of the board as the cut made in step 6.
10. Repeat steps 7–9 three times to cut the remaining three legs.

Figure 51
Figure 51 illustrates step 3

Figure 52
Figure 52 illustrates step 7

Figure 53
Figure 53 illustrates step 8

Lay Out and Cut the Top

The top of the sawhorse will be cut from a piece of dimensional lumber using a circular saw. First the board must be cut to a length of 4′-0″. To produce the bevel, a 12° cut will be made along the length of the board on both sides.

Figure 54
Figure 54 illustrates step 2

PROCEDURE

1. Measure, mark, and cut the 2 x 6 top board to a length of 4′-0″.
2. Measure back 6″ from each end and square a line across the width of the top face of the top board. See Figure 54.
3. Set the base plate of the circular saw to an angle of 12°, and cut a 12° bevel the length of both sides of the top board, maintaining a 5 ½″ width on one face of the top board. See Figure 55. Move the clamps along the length of the board as necessary.
4. Set the sliding T-bevel at the angle used to lay out the legs.
5. Mark a line on each edge of the board at the marks made in step 2 using the T-bevel set to the angle in step 4. See Figure 56.

Figure 55
Figure 55 illustrates step 3

Figure 56
Figure 56 illustrates step 5

Attach the Legs, Gussets, and Side Spreaders

Attaching the legs, gussets, and side spreaders is a task requiring more than 20 individual steps, each of which must be carefully completed. When attaching the legs, make sure the top of each leg is flush with the top board. The gussets will be laid out on plywood using a framing square and sliding T-bevel. Once cut, the gussets are attached with construction screws. The side spreaders will be cut from 1 x 4 lumber.

Figure 57
Figure 57 illustrates step 1

PROCEDURE

1. Drill four pilot holes parallel to the top end of each leg. Drill the holes ¾″ deep. See Figure 57.
2. Align one leg with an angled line drawn on the beveled edge of the top board. Be sure that the leg top is flush with the top of the top board. See Figure 58.

Figure 58
Figure 58 illustrates step 2

attach the legs

Figure 59
Figure 59 illustrates step 3

Attach the leg to the top board...

3. Attach the leg to the top board with four 1 ⅝″ construction screws. See Figure 59.
4. Repeat steps 2 and 3 for the remaining legs. See Figure 60.
5. Cut two pieces of ½″ plywood 12″ x 15″ to make the gussets.
6. Place the plywood pieces against the bottom of the top board.
7. Mark the outside edge of the legs where the plywood pieces meet the bottom of the top board. See Figure 61.

Figure 60
Figure 60 illustrates step 4

Figure 61
Figure 61 illustrates step 7

8. Align the 12″ mark on the body of the framing square and the 2 ⅝″ mark on the tongue of the framing square with the edge of a piece of scrap material and draw a line along the tongue. See Figure 62.
9. Set the sliding T-bevel to the line drawn in step 8. See Figure 63.
10. Draw angled lines, using the sliding T-bevel, from the top marks made in step 7 to the bottom of the plywood pieces. Extend the lines with a straightedge.
11. Cut along the angled lines drawn on the plywood with the circular saw to complete two gussets.
12. Place the sawhorse upside down on a flat surface and set the spread of the legs to19 ⅛″ at each end.
13. Clamp a piece of scrap to the legs to hold them in place.

Figure 62
Figure 62 illustrates step 8

Figure 63
Figure 63 illustrates step 9

Figure 64
Figure 64 illustrates step 14

Figure 65
Figure 65 illustrates step 15

14. Place a plywood gusset on one end under the top board and adjust the legs to align with the edges of the gusset. Clamp the gusset to the legs and remove the scrap piece. See Figure 64.
15. Fasten the gusset to the legs with eight 1 ¼″ construction screws. See Figure 65.
16. Repeat steps 13 and 14 for the other end of the horse to complete the gusset assembly.
17. Cut the 1 x 4 x 8′-0″ in half, using a circular saw. The resulting pieces will be used for the two side spreaders.
18. Set the spread of the legs at the bottom to 4′-0″ and position one piece of the 1 x 4 cut in step 17 against the two side legs of the horse. Align the bottom of the 1 x 4 with the bottom of the plywood gussets. See Figure 66.

Figure 66
Figure 66 illustrates step 18

19. Draw lines on the back of the 1 x 4 along the edge of the gussets to mark a side spreader. See Figure 67.
20. Cut the side spreader to length with the circular saw along the lines drawn in step 17.
21. Attach the side spreader to the legs with four 1 5⁄8″ construction screws. Use two per leg. See Figure 68.
22. Repeat steps 17–21 for the second side spreader.
23. Check the sawhorse for wobble on a flat surface and adjust the leg length as needed.

Figure 67
Figure 67 illustrates step 19

Check the sawhorse for wobble on a flat surface...

Figure 68
Figure 68 illustrates step 21

Student Name: ______________________________ Date: ______________

Notched-Top Sawhorse Project Evaluation

PROCEDURE		CRITERIA AS SPECIFIED BY THE PROJECT	POSSIBLE POINTS	SCORE
Cut the Pieces/Lay Out and Cut Notches on the End Pieces				
Horse	**Legs (4)**	Length	10	
		Angle	10	
		Bevel	10	
		Subtotal	*30*	
	Top	Length	5	
		Notch Location	5	
		Angle of notches (4)	10	
		Depth of notches (4)	10	
		Subtotal	*30*	
	Gussets (2)	Cut at correct angle	5	
		Cut to correct dimensions	5	
		Subtotal	*10*	
	Spreaders (2)	Cut at correct angle	5	
		Cut at correct length	5	
		Subtotal	*10*	
Assembly	Overall height		5	
	Spread of sawhorse legs at base on sides		5	
	Spread of sawhorse legs at base on ends		5	
	Fasteners located as per drawing		5	
	Splits and shiners		5	
	Edges flush		5	
	Joints tight		5	
		Subtotal	*35*	
General	Tool handling		5	
	Followed direction		5	
	Cleaned up		5	
	Safe work practices		5	
		Subtotal	*20*	
			Student score:	

Total Possible Points = 135

Suggested minimum acceptable score: 94 points or ______

Student's Signature: ______________________ Teacher's Signature: ______________________

Student Name:________________________________ Date: ________________

I-Beam Sawhorse Project Evaluation

PROCEDURE		CRITERIA AS SPECIFIED BY THE PROJECT	POSSIBLE POINTS	SCORE
Cut the Pieces/Lay Out and Cut Notches On the End Pieces				
Horse	**Legs (4)**	Length	10	
		Angle	10	
		Bevel	10	
		Subtotal	*30*	
	Top	Length	5	
		Attached properly to I-Beam	5	
		Subtotal	*10*	
	I-Beam	Length	5	
		Assembled per procedures	5	
		Location of legs marked properly	5	
		Subtotal	*15*	
	Gussets (2)	Cut at correct angle	5	
		Cut to correct dimensions	5	
		Subtotal	*10*	
Assembly	Fasteners located as per drawing		5	
	Overall height		5	
	Spread of sawhorse legs at base on sides		5	
	Spread of sawhorse legs at base on ends		5	
	No splits and shiners		5	
	Edges flush		5	
	Joints tight		5	
		Subtotal	*35*	
General	Tool handling		5	
	Followed direction		5	
	Cleaned up		5	
	Safe work practices		5	
		Subtotal	*20*	
			Student score:	

Total Possible Points = 120

Suggested minimum acceptable score: 84 points or ______

Student's Signature: ____________________ Teacher's Signature: ____________________

Student Name: ______________________________ Date: ______________

Beveled-Top Sawhorse Project Evaluation

PROCEDURE		CRITERIA AS SPECIFIED BY THE PROJECT	POSSIBLE POINTS	SCORE
Cut the Pieces/Lay Out and Cut Notches on the End Pieces				
Horse	**Legs (4)**	Length	10	
		Angle	10	
		Bevel	10	
		Subtotal	*30*	
	Top	Length	5	
		Leg Location	5	
		Subtotal	*10*	
	Gussets (2)	Cut at correct angle	5	
		Cut to correct dimensions	5	
		Subtotal	*10*	
	Spreaders (2)	Cut at correct angle	5	
		Cut at correct length	5	
		Subtotal	*10*	
Assembly	Overall height		5	
	Spread of sawhorse legs at base on sides		5	
	Spread of sawhorse legs at base on ends		5	
	Fasteners located as per drawing		5	
	No splits and shiners		5	
	Edges flush		5	
	Joints tight		5	
		Subtotal	*35*	
General	Tool handling		5	
	Followed direction		5	
	Cleaned up		5	
	Safe work practices		5	
		Subtotal	*20*	
			Student score:	

Total Possible Points = 115

Suggested minimum acceptable score: 80 points or ______

Student's Signature: ______________________ Teacher's Signature: ______________________

CHAPTER 4

PICNIC TABLE

CONTENTS

Introduction

This chapter contains procedures for building three different kinds of picnic tables. Section one of this chapter is devoted to construction of a curved picnic table. Section two explains how to build a folding picnic table, while section three will enable you to build an octagonal picnic table. The processes involved in construction of these three types of tables are similar in some ways and quite different in others. Regardless of which one you build, you will learn several more construction techniques that are used by professional carpenters.

In the building of a picnic table, you will get additional practice working with common carpentry materials such as pine and cedar lumber and fasteners such as screws, nuts, bolts, and washers. You will also get an opportunity to develop more skill using important carpentry tools such as a Speed® Square, compass, circular saw, saber saw, belt sander, and router. You may even get a chance to put to good use the sawhorse you built in the previous chapter.

What's New?

The projects in this chapter are intended to expand your carpentry knowledge and skills. Completing a curved, folding, or octagonal picnic table will help you learn about several commonly used carpentry procedures, tools, and fasteners. Among these are the following:

Carriage Bolt Bolts are threaded metal fasteners often used in conjunction with a nut and washer. Carriage bolts have round heads and a square neck that prevents the bolt from turning while the nut is tightened. Figure 1 illustrates a carriage bolt.

Figure 1
Carriage bolt

Countersink When carpenters don't want the heads of screws or bolts to extend above a surface, they countersink them. This means that they use a drill bit or a countersink bit to make a depression in the material, shown in Figure 2, so that when the bolt or screw is tightened, the head will be flush with or below the surface.

Deck Screw A deck screw is similar to a construction screw but has been treated with chemicals or coatings to make it more resistant to corrosion. Deck screws are shown in Figure 3. Because of their corrosion resistance, deck screws are often used outdoors where fasteners and materials are likely to be exposed to weather and high levels of moisture.

Figure 2
Countersink for a screw

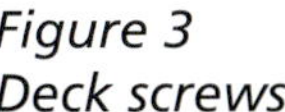

Figure 3
Deck screws

Figure 4
Roundover bit

Roundover Bit Among the many bits available for use with routers, the roundover bit produces a simple, rounded edge. A roundover bit is shown in Figure 4. It can also cut an edge with a stepped shoulder.

Strap Hinge Like most other hinges, strap hinges are made of metal and are used for attaching components that must swing open and shut. An example is shown in Figure 5. Strap hinges are not typically recessed. Instead they are mounted on the surface.

Figure 5
Strap hinge

1 Curved Picnic Table

Because of the curved top and benches, this picnic table is challenging to build. Its construction requires measuring and laying out both concave and convex curves, which must be uniform from board to board. This is why the procedures use one component as a pattern to lay out other curved parts to make the job easier and more precise. Also, the actual cutting of the curves requires careful attention to detail and a steady hand. And finally, even the assembly stage is complex because the curved boards must be fastened slightly differently. Although complex, the 16 separate procedures that make up this section will provide detailed guidance in making this attractive and functional picnic table.

...the procedures include the making of templates...

You will need the following materials:

- (9) 2 x 4 x 8′-0″ cedar lumber
- (1) 2 x 6 x 10′-0″ cedar lumber
- (6) 5⁄4 x 6 x 12′-0″ cedar deck boards
- (1) 1 x 2 x 8′-0″ pine lumber
- (1) 1 x 4 x 8′-0″ pine lumber
- (36) 3⁄8″ x 3″ galvanized carriage bolts
- (36) 3⁄8″ galvanized nuts and washers
- (5 lb) 2 1⁄2″ deck screws
- (1 lb) 1 1⁄4″ construction screws

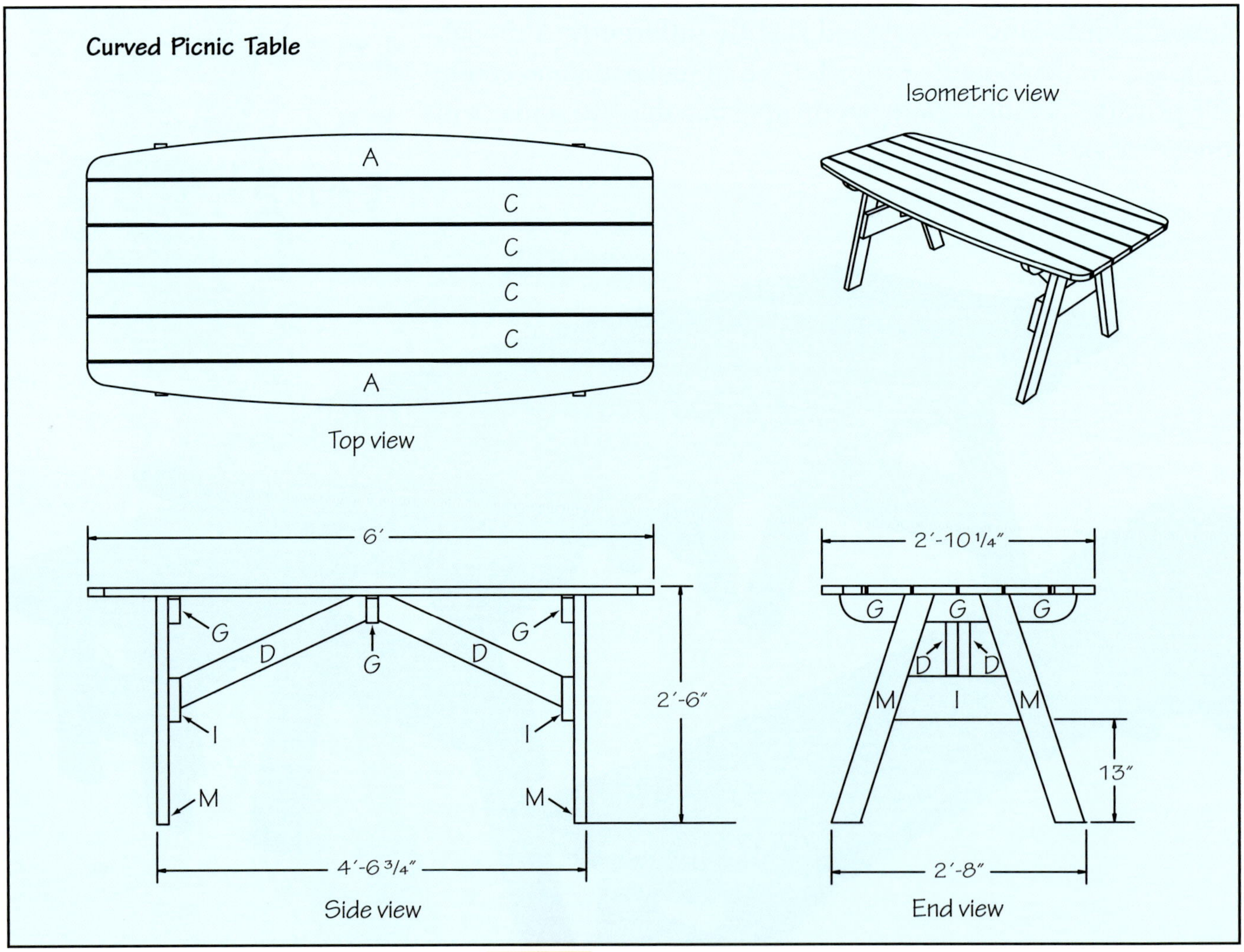

Curved Picnic Table

A
C
C
C
C
A

Top view

Isometric view

6′
G
D
G
D
G
2′-6″
I
I
M
M
4′-6 3/4″

Side view

2′-10 1/4″
G
G
G
D
D
M
I
M
13″
2′-8″

End view

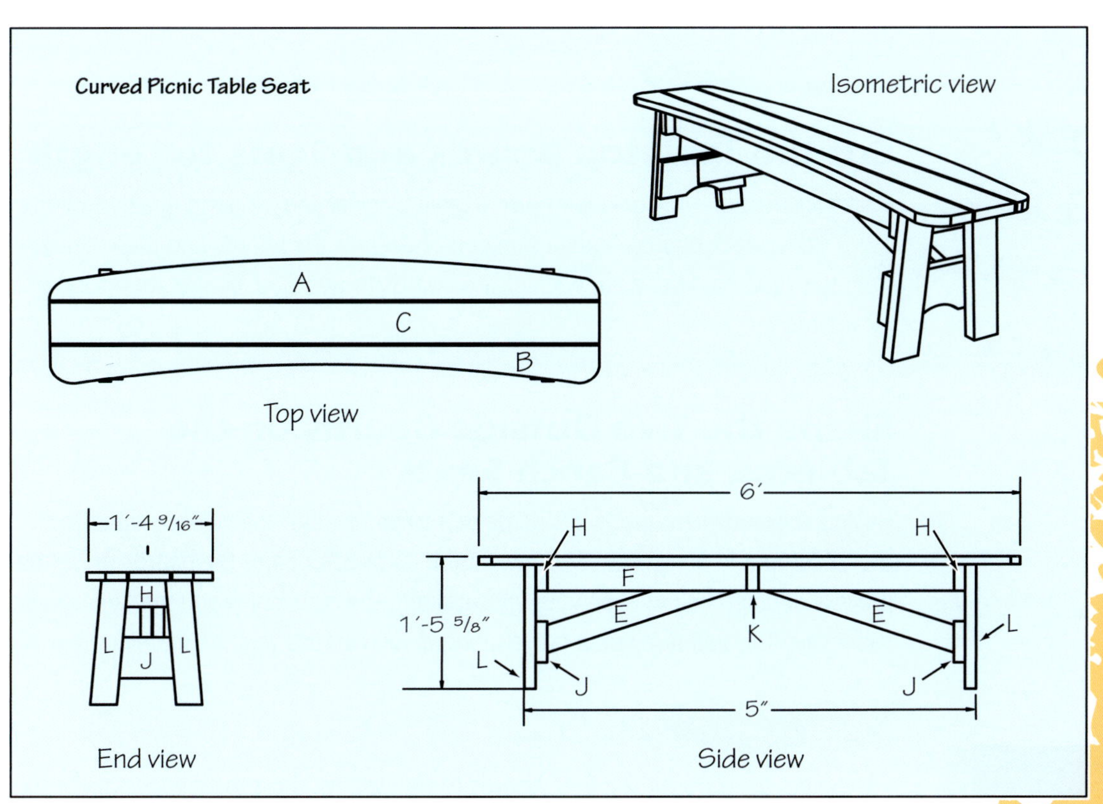

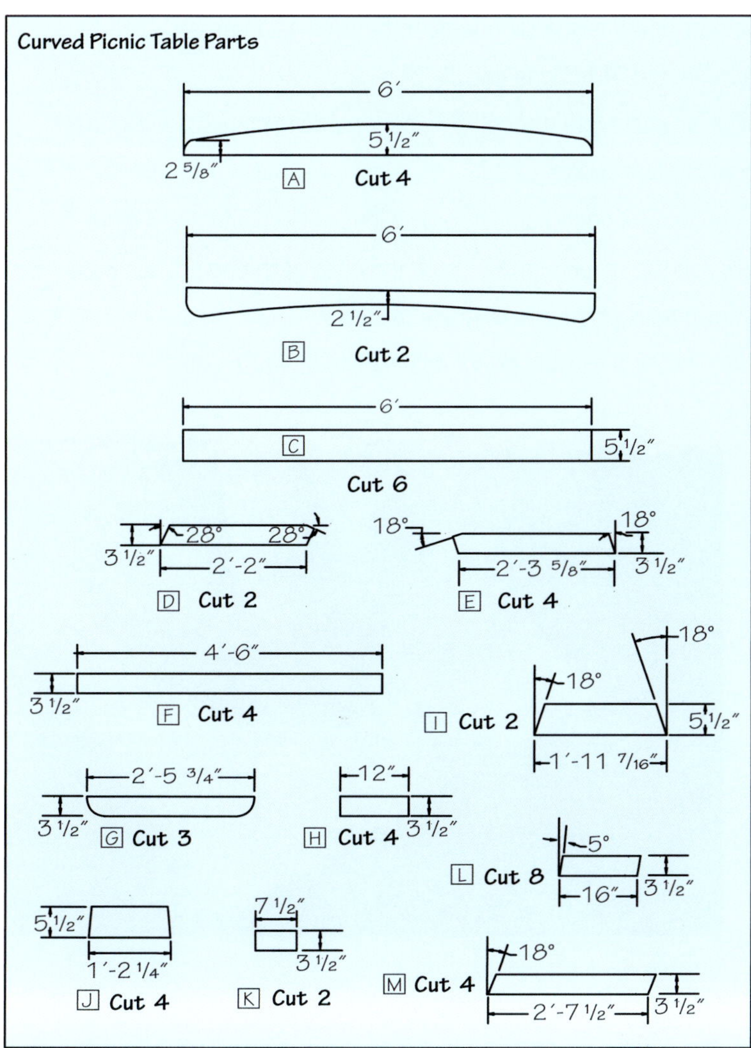

You will need the following tools:

- Circular saw
- Clamps
- Compass
- Random orbital sander
- Straightedge
- Tape measure
- 3⁄16″ roundover bit
- Router
- Saber saw
- Sanding block
- Sandpaper 120 grit and 220 grit
- (2) Sawhorses
- Screw gun
- Sliding T-bevel
- 1″ spade bit
- Speed square
- Socket wrench and sockets
- Electric drill
- Drill index

. . . cut along the curved line you marked using a saber saw. . .

Cut the Tabletop Boards and Seats to Length

Start construction of your curved picnic table by selecting six good quality 5/4 x 6 x 12′-0″ cedar deck boards. Square one end of each of the boards and then measure from the square end and cut each of the boards to two 6′-0″ boards for a total of 12 boards.

Shape the Two Outside Boards on the Tabletop and Bench Seats

Having selected six of the 5/4 x 6′-0″ boards to be used for the curved edges of the tabletop and bench seats, you will mark an arc on one of the boards using the process outlined below. Then you will cut along the curved line you marked using a saber saw. This will give you a pattern board you can use to cut remaining boards.

PROCEDURE

1. Select the six best 5/4 x 6′-0″ boards to be used for the curved edges of the top and the benches.
2. Select the best board of the six boards to be used as a pattern.
3. Place a mark at 3′-0″ from one end.
4. Lightly draw a square line through the mark made in step 3. See Figure 6.
5. Set the combination square to 2 ½″.
6. Scribe a line about 2″ long parallel to the sides at each end. See Figure 7.
7. Place marks on the scribed lines 1″ from the ends of the board. See Figure 8.
8. Drive 8d finish nails at the marks made in step 7.
9. Position the 1 x 2 x 8′-0″ pine board against the 8d finish nails.

Figure 6
Figure 6 illustrates step 4

Figure 7
Figure 7 illustrates step 6

Figure 8
Figure 8 illustrates step 7

Figure 9
Figure 9 illustrates step 10

Figure 10
Figure 10 illustrates step 13

10. Spring the 1 x 2 board towards the edge of the board at the 3′-0″ centerline. When the edge of the 1 x 2 is even with the edge of the 5/4 x 6 board, the desired arc has been created. See Figure 9.
11. Trace the arc on the 5/4 x 6 while maintaining the 1 x 2 in the position described in step 10.
12. Take the 1 x 2 off and remove the finish nails.
13. Measure and place a mark 2 ¾″ in from one end of the 5/4 x 6 board. See Figure 10.
14. Lightly draw a square line through the mark. See Figure 11.
15. Set the compass to a radius of 2 ¾″.

Figure 11
Figure 11 illustrates step 14

Figure 12
Figure 12 illustrates step 16

16. Position the compass where the squared line meets the squared edge on what will be the straight edge of the board. See Figure 12.
17. Swing an arc with the compass from the corner to where it intersects the curved line. See Figure 13.
18. Repeat steps 13–17 at the other end of the board.
19. Cut the curved line made in step 11 and the two radius corners, using the saber saw.
20. Sand any irregularities on the cuts to smooth the edges, if necessary. The pattern board is now complete. See Figure 14.
21. Use the pattern board to make three more identical curved boards from the other five best boards – two for the bench and one for the top.
22. Sand any irregularities on the cuts to smooth the edges, if necessary.

Figure 13
Figure 13 illustrates step 17

Figure 14
Figure 14 illustrates step 20

Make the Concave Cut for Bench Tops

The inside edge of the bench tops will be slightly concave to make them more comfortable. To make them concave you will mark and cut along a curved line just as you did in the previous procedure. Once you have marked and cut a pattern board you can use it to cut a second concave seat piece.

Figure 15
Figure 15 illustrates step 3

PROCEDURE

1. Select one of the boards from the remaining two best 5/4 x 6 boards.
2. Make a mark 3′-0″ from one end of the 5/4 x 6 board.
3. Lightly draw a square line through the mark made in step 2. See Figure 15.
4. Place a mark 3″ from one edge on this line. See Figure 16.
5. Place the pattern created in the previous procedure aligning the curve with the 3″ mark at the centerline and the ends. Note: if the pattern rocks on the surface, slide another 5/4 x 6 beside the board to support the pattern as shown in Figure 17.
6. Trace the curve from the pattern to the bench top.
7. Set the pattern aside.

Figure 16
Figure 16 illustrates step 4

Trace the curve from the pattern...

Figure 17
Figure 17 illustrates step 5

Figure 18
Figure 18 illustrates step 8

8. Measure from the end 2 ¾″ and square the line across. See Figure 18.
9. Measure down the line 2 ¾″ and place a mark. See Figure 19.
10. Set the compass to a radius of 2 ¾″.
11. Swing an arc with the compass positioned at the mark established in step 9. See Figure 20.
12. Repeat steps 8–11 for the other end of the board.
13. Cut the curved line and the two radius corners laid out in this procedure, using the saber saw.

Figure 19
Figure 19 illustrates step 9

Figure 20
Figure 20 illustrates step 11

Swing an arc with the compass...

Figure 21
Figure 21 illustrates step 14

14. Sand any irregularities on the cuts to smooth the edges, if necessary. See Figure 21.
15. Use this board as a pattern to mark the second concave bench top board.
16. Repeat steps 13 and 14 to complete the shaping of the second board.
17. Round over all the cut edges of the 5⁄4 x 6 boards, using the router with a 3⁄16″ radius roundover bit. Be sure to roundover the cross cut ends and not just the curved cuts. See Figure 22.

Figure 22
Figure 22 illustrates step 17

Cut the Supports for the Tabletop

Cutting the supports for the tabletop involves marking and cutting along curved lines. You will use a compass to create the arc and a saber saw to make the cut. Having created a pattern board, you can use it to cut the remaining two supports.

PROCEDURE

1. Cut three 2 x 4 x 2′-5 ¾″ pieces, from one 2 x 4 x 8′-0″, for the tabletop supports.
2. Measure back 3 ½″ along an edge from each end on one of the pieces from step 1 and make a mark. See Figure 23.
3. Set the compass to a radius of 3 ½″.
4. Set the point of the compass to the mark established in step 2 and swing an arc at the end of the board. See Figure 24.
5. Repeat step 4 at the other end.
6. Cut the curved lines with saber saw. Sand and smooth as necessary.
7. Use this piece as a pattern to mark and cut the other two pieces made in step 1.
8. Sand and smooth as necessary. See Figure 25.

Figure 23
Figure 23 illustrates step 2

Figure 24
Figure 24 illustrates step 4

Figure 25
Figure 25 illustrates step 8

Cut the Legs for the Table

When cutting the legs for the table, you will be working with angles rather than arcs. As you have done in previous procedures, you will mark and cut one leg and use that as a pattern for cutting the remaining legs. The pieces of lumber left over from the cuts will be set aside for use as braces.

Figure 26
Figure 26 illustrates step 1

PROCEDURE

1. Mark and cut one end of the 2 x 4 x 8′-0″ at an 18° angle. See Figure 26.
2. Measure 2′-7 ½″ from the long point of the cut end as shown in Figure 27a, and cut a parallel 18° angle as shown in Figure 27b. This will be the pattern for the other three legs.

Figure 27
(a) Figure 27 illustrates step 2

(b) Figure 27 illustrates step 2

Use the pattern to mark and cut the second leg...

3. Use the pattern to mark and cut the second leg from the same 2 x 4. Set the remaining piece aside to be used later.
4. On another 2 x 4 x 8′-0″, use the pattern to mark and cut the remaining two legs. See Figure 28. Set the remaining piece aside to be used later.

Figure 28
Figure 28 illustrates step 4

28 degrees

Figure 29
Figure 29 illustrates completed step 2

Cut the Table Angle Braces

You will use the two pieces of cut lumber left over from the previous procedure to mark and cut angle braces for the table. Here you will be working with angles of 18° and 28°. The first completed brace can be used as a pattern for cutting the second one.

PROCEDURE

1. Choose one of the two remaining pieces left over from the legs.
2. Mark a 28° angle across the face of the board using the long point of the 18° angle as the long point of the 28° angle. See Figure 29. Cut the board along the line.
3. Measure and mark 2′-2″ along the edge of the board from the long point of the 28° angle made in step 2. See Figure 30.
4. Cut a 28° angle parallel to the angle cut at the other end of the board.

Figure 30
Figure 30 illustrates step 3

Figure 31
Figure 31 illustrates step 5

5. Measure back 1 ½″ along the edge of the board, from the long point of the angle cut in step 4 and place a mark. See Figure 31.
6. Draw a line through the mark square to the 28° angle cut. See Figure 32.
7. Cut along the line laid out in step 6. This will be the end of the brace that attaches to the center support.
8. Use this piece as a pattern to mark and cut the second table brace from the other piece of leftover material from the legs. See Figure 33.

Figure 32
Figure 32 illustrates step 6

Figure 33
Figure 33 illustrates step 8

Cut the Spreaders for the Legs

Spreaders are supports that help hold legs or other members a proper distance apart. The spreaders for the picnic table legs are cut from cedar lumber at an 18° angle. The first spreader will be used as a pattern for the second one, and the remaining lumber will be set aside for later use as bench spreaders.

PROCEDURE

1. Cut an 18° angle on one end of the 2 x 6 x 10′-0″.
2. Measure 23 ½″ from the long point of the angle, mark and cut an opposing 18° angle. See Figure 34.
3. Align the edges of the 18° angle cuts, using the piece cut in step 2 as a pattern. See Figure 35. Mark and cut the second leg spreader. Set what is left of the 2 x 6 aside for bench spreaders.

Figure 34
Figure 34 illustrates step 2

Figure 35
Figure 35 illustrates step 3

Assemble the Legs

Assembling the legs of the picnic table is an involved process requiring very careful attention to detail. You will create two separate leg assemblies. Keep in mind that the proper alignment of the legs is essential. For appearance's sake you will want to countersink some of the carriage bolts so they will be recessed below the surface.

PROCEDURE

1. Place a mark at the centerline of the tabletop support that was completed in the "Cut the Supports for the Tabletop" procedure.
2. Place a mark 3″ and 4 ¾″ on each side of the center mark.
3. Draw an 18° line through each mark. See Figure 36.
4. Measure down 1 ¾″ on the 4 ¾″ mark made in step 2 along the 18° angle line drawn in step 3 and place a mark. See Figure 37.
5. Drill a 1″ diameter countersink hole ⅜″ deep at the mark laid out in step 4. See Figure 38.
6. Repeat steps 1–5 for the second tabletop support. The third support will not require countersink holes.

Figure 36
Figure 36 illustrates step 3

Figure 37
Figure 37 illustrates step 4

Figure 38
Figure 38 illustrates step 5

Figure 39
Figure 39 illustrates step 7

7. Measure in 1 ¾″ from each end of one spreader and place a mark. See Figure 39.
8. Draw an 18° line through each mark parallel to end of the spreader. See Figure 40.
9. Place marks 1 ¼″ from the top and bottom of the spreader along the 18° lines drawn in step 8. See Figure 41.
10. Drill a 1″ diameter countersink hole ⅜″ deep at the marks laid out in step 9.

Figure 40
Figure 40 illustrates step 8

Figure 41
Figure 41 illustrates step 9

Figure 42
Spreaders with countersink holes

11. Repeat steps 7–10 for the second spreader. Spreaders with countersink holes are shown in Figure 42.
12. Lay the two legs on a flat surface and align the top of the support with the top of the legs.
13. Align the inside edge of the legs with the lines drawn on the support through the 3″ marks made in step 2. Make sure to maintain the alignment from step 12.
14. Secure this alignment with two clamps, one on each leg. See Figure 43.

Figure 43
Figure 43 illustrates step 14

Drill two 3/8" diameter holes through the assembly...

Figure 44
Figure 44 illustrates step 15

15. Drill two ⅜″ diameter holes through the assembly using the center points of the countersink holes as guides. See Figure 44.
16. Secure the support to the legs with two ⅜″ x 3″ carriage bolts, nuts, and washers, positioning the carriage bolt head against the face of legs and the nuts and washers in the countersink holes. See Figure 45.
17. Lay the spreader on top of the legs.
18. Put the straight edge on the bottom of the legs.

Figure 45
Figure 45 illustrates step 16

Figure 46
Figure 46 illustrates step 19

19. Position the spreader parallel to the straight edge 13″ away. Align the ends of the spreader with the edge of the legs. See Figure 46.
20. Clamp the spreader to the legs.
21. Drill four ⅜″ diameter holes through the assembly using the center points of the countersink holes as guides. See Figure 47.
22. Secure the spreader to the legs with four ⅜″ x 3″ carriage bolts, nuts, and washers, positioning the carriage bolt head against the face of legs and the nuts and washers in the countersink holes. A completed leg assembly is shown in Figure 48.
23. Repeat steps 1–22 to assemble the second leg assembly.

Figure 47
Figure 47 illustrates step 21

Figure 48
Completed leg assembly

Assemble the Tabletop

While assembling the tabletop, you will temporarily secure the various pieces with 4′-0″ cleats cut from a piece of pine lumber. One of the cleats will be placed ¾″ from the center of the table and will help you position the center support. Once in position, the center support will be secured to the tabletop with 12 corrosion resistant deck screws. Later the cleats will be removed.

Lay the tabletop boards face down on a flat surface...

PROCEDURE

1. Lay the tabletop boards face down on a flat surface approximately ¼″ apart, with the two curved pieces on the outside and ends of the boards aligned. See Figure 49.
2. Cut the 1 x 4 x 8′ pine board in two pieces to make two temporary cleats approximately 4′-0″ long.
3. Position one cleat over the tabletop boards roughly 6″ away from an end. Locate the center of the table and position the second cleat ¾″ off center. See Figure 50.
4. Temporarily fasten the cleats to the tabletop boards with 1 ¼″ construction screws.
5. Turn the tabletop over on the flat surface.
6. Slide two 2 x 4s on edge underneath each end of the tabletop, avoiding the cleats.

Figure 49
Figure 49 shows step 1 completed

Figure 50
Figure 50 illustrates step 3

Figure 51
Figure 51 illustrates step 7

Figure 52
Figure 52 illustrates step 8

7. Locate the center of the tabletop and draw a line from edge to edge. See Figure 51.
8. Place marks along the line 1″ in from the edges of each board. See Figure 52.
9. Drill the appropriate size pilot holes at each mark.
10. Slide the center support underneath the tabletop against the center cleat.
11. Center this support between the curved sides of the table.
12. Secure the tabletop to the center support with twelve 2 ½″ deck screws, two in each board. See Figure 53.

Figure 53
Figure 53 illustrates step 12

Attach the Braces and Legs

Now it's time to attach the braces and legs to the tabletop. The temporary center cleat will be removed and the braces will be attached to the center support by inserting deck screws into pilot holes. Then each of the leg assemblies are positioned with the help of a temporary cleat and securely attached to the tabletop with ten 2 ½″ deck screws.

Secure the braces to the center support...

PROCEDURE

1. Turn the tabletop over on the flat surface.
2. Remove the temporary center cleat.
3. Draw a centerline on both faces of the center support.
4. Drill four pilot holes, two on each side of the centerline. See Figure 54. Drill the holes ¾″ away from the centerline.
5. Butt the end of one brace to the center support on one side of the centerline.
6. Secure the braces to the center support with 2 ½″ deck screws. See Figure 55.

Figure 54
Figure 54 illustrates step 4

Figure 55
Figure 55 illustrates step 6

Figure 56
Figure 56 illustrates step 10

7. Remove the other temporary cleat.
8. Draw a centerline on the spreader of one leg assembly.
9. Test fit one leg assembly by placing it on the tabletop and sliding it to the brace.
10. Align the mark on the spreader of the leg assembly to the brace connected to the center support. Make sure the brace is on the same side of the centerline as it is attached to the center table support. See Figure 56.
11. Locate and drill two pilot holes through the spreader.
12. Secure the spreader to the brace with 2 ½″ deck screws. See Figure 57.

Figure 57
Figure 57 illustrates step 12

Square the leg assembly to the tabletop.

Figure 58
Figure 58 illustrates step 13

13. Square the leg assembly to the tabletop. See Figure 58.
14. Position and secure a temporary cleat to the tabletop next to the leg assembly, maintaining the spacing of the tabletop boards.
15. Center the leg assembly between the curved sides of the tabletop.
16. Clamp the leg assembly to the temporary cleat. See Figure 59.

Figure 59
Figure 59 illustrates step 16

Figure 60
Figure 60 illustrates step 17

17. Repeat steps 8–16 for the other leg assembly. See Figure 60.
18. Turn the table over on the flat surface.
19. Locate the centerline of the leg assemblies and draw a line on the tabletop from edge to edge. See Figure 61.
20. Center the support between the sides of the table.
21. Place marks along the line 1″ in from the edges of each board. The curved boards on the sides of table will only have one mark in each end. The mark should be centered. See Figure 62.
22. Drill the appropriate size pilot holes at each mark.
23. Secure the tabletop to the leg assemblies with ten 2 ½″ deck screws, two in each board. The two curved boards on the sides will have only one screw in each end.

Figure 61
Figure 61 illustrates step 19

Figure 62
Figure 62 illustrates step 21

Cut Parts for the Benches

The picnic table benches consist of a number of parts that must be carefully measured and cut for both benches. These parts include a seat frame center support, seat frame ends, seat frame sides, bench legs, bench leg spreaders, and braces.

PROCEDURE

1. Cut one 4′ -6″ seat frame side and one 12″ seat frame end from one 2 x 4 x 8′. Set the remainder aside for one bench angle brace.
2. Repeat step 1 on three more 2 x 4 x 8′.
3. Cut five 16 ½″ legs with a 5° parallel angle at both ends and one 7 ½″ seat frame center support with two square ends, from one 2 x 4 x 8′. See Figure 63.

Figure 63
Figure 63 illustrates step 3

4. Cut three 16 ½″ legs with a 5° parallel angle at both ends and one 7 ½″ seat frame center support with two square ends, from one 2 x 4 x 8′.
5. Cut four bench leg spreaders 14 ¼″ long, with opposing 5° angles at the ends, from the remaining 2 x 6 left over from the table legs. The spreaders are 14 ¼″ from long point to long point. See Figure 64.

Figure 64
Figure 64 illustrates step 5

Figure 65
Figure 65 illustrates step 6

6. Cut four 2′-3 ⅝″ bench angle braces with an 18° parallel angle at each end, from the pieces set aside in steps 1 and 2. See Figure 65.
7. Measure back 1 ¾″ along the edge of the board, from the long point of the angle and place a mark. See Figure 66.

Figure 66
Figure 66 illustrates step 7

8. Draw a line through the mark square to the angle cut. See Figure 67.
9. Cut along the line from step 8. This will be the end of the brace that attaches to the bench support.
10. Use this piece as a pattern to mark and cut the three other bench angle braces. See Figure 68.

Figure 67
Figure 67 illustrates step 8

Figure 68
Figure 68 illustrates step 10

Figure 69
Figure 69 illustrates step 1

Assemble the Upper Frame of the Benches

Now that the parts have been cut, the upper frame of both benches can be assembled. Parts will be secured using deck screws inserted into pilot holes. When drilling pilot holes through the seat frame sides, be careful not to drill into the center support.

PROCEDURE

1. Drill four pilot holes in the 2 x 4 x 12″ seat frame end, two holes ¾″ from one end and another two holes 9 ¾″ measured from the same end. See Figure 69.
2. Repeat step 1 for the other three 2 x 4 x 12″ seat frame ends.
3. Attach one 2 x 4 x 12″ seat frame end to each end of the 2 x 4 x 4′-6″ seat frame side with two 2 ½″ deck screws through each end. See Figure 70.

Figure 70
Figure 70 illustrates step 3

4. Make a mark at the center of the 7 ½" seat frame center support on the edge at 3 ¾". See Figure 71.
5. Use this seat frame center support as a temporary spacer by positioning and clamping it with the end butted up to the seat frame side and the face to the seat frame end.
6. Butt the other 2 x 4 x 4′-6″ seat frame side against the seat frame center support and secure the seat frame end to the side with one 2 ½″ deck screw. See Figure 72.

Figure 71
Figure 71 illustrates step 4

Figure 72
Figure 72 illustrates step 6

Figure 73
Figure 73 illustrates step 7

Figure 74
Figure 74 illustrates step 10

7. Transfer the center mark made in step 4 on the 7 ½″ seat frame center support to the 12″ seat frame end. See Figure 73.
8. Repeat steps 5–7 for the other end.
9. Measure and mark the center of the 4′-6″ seat frame side pieces.
10. Center the 2 x 4 x 7 ½″ seat frame center support on the marks made in step 9. See Figure 74.
11. Drill two pilot holes through both seat frame sides for the center support. Be careful not to drill into the support.

Drill two pilot holes through both seat frame sides...

Figure 75
Figure 75 illustrates step 12

12. Secure the center support in place with four 2 ½″ deck screws, two through each seat frame side piece. See Figure 75.
13. Install 2 ½″ deck screws in all of the remaining pilot holes.
14. Repeat steps 3–13 for the second frame. Completed upper seat frames are shown in Figure 76.

Figure 76
Completed upper frames

Assemble the Bench Legs

The leg assemblies of each bench are put together by attaching the legs and spreaders. A piece of 2 x 4 lumber will be used as a spacer to help position the spreaders. This procedure must be repeated four times, once for each leg assembly.

PROCEDURE

1. Lay two 2 x 4 x 16 ½″ legs on a flat surface.
2. Place one 2 x 6 x 14 ¼″ bench leg spreader on top of the legs.
3. Place a 2 x 4 of sufficient length on top of and across the bottom of both legs. Align the bottoms of the legs flush with the bottom edge of the 2 x 4. The 2 x 4 will be used as a spacer to position the spreader.
4. Move the spreader until it touches the top of the 2 x 4 spacer.
5. Adjust the legs until the outside edge aligns with outside edges of the spreader. See Figure 77. Keep the spreader and the 2 x 4 spacer in alignment while adjusting the legs.
6. Drill pilot holes. Secure the spreader to the legs with one 2 ½″ deck screw in each leg 1″ from the outside edge and 1″ from the bottom of the spreader. See Figure 78.
7. Repeat steps 1–6 for the remaining three sets of legs.

Figure 77
Figure 77 illustrates step 5

Figure 78
Figure 78 illustrates step 6

Attach the Bench Legs to the Bench Frame

The completed bench leg assemblies will now be attached to the bench frame. Once the corners of the bench leg assemblies and frames are properly aligned, they will be securely clamped. Then pilot holes will be drilled for the deck screws that will hold these pieces together. When the combined assemblies are completed they will be turned right side up on a flat surface to make sure each of the legs makes stable contact with the surface.

Figure 79
Figure 79 shows steps 1 and 2 completed

PROCEDURE

1. Position one leg assembly upside down and against the end of the frame assembly with the spreader on the inside of the leg.
2. Align the corners of the leg assembly with the corners of the frame assembly and clamp in place. See Figure 79.
3. Fasten one completed leg assembly to the frame with one 2 ½" construction screw in each leg. See Figure 80.

Figure 80
Figure 80 illustrates step 3

Figure 81
Figure 81 illustrates step 5

Figure 82
Figure 82 illustrates step 6

4. Repeat steps 1– 3 to fasten another completed leg assembly to the other end of the frame assembly.
5. Draw a centerline on both faces of the center support. See Figure 81.
6. Transfer the centerlines marked on the seat frame ends in step 7 of the procedure "Assemble the Upper Frame of the Benches" to the leg spreaders. See Figure 82.
7. Drill four pilot holes in the center support for the braces, two ¾″ away on each side of the centerline. See Figure 83.
8. Butt the end of one brace to the center support on one side of the centerline.
9. Secure the brace to the center support through the pilot holes with 2 ½″ deck screws.

Figure 83
Figure 83 illustrates step 7

Figure 84
Figure 84 illustrates step 10 completed

10. Repeat steps 8 and 9 for the other brace. See Figure 84.
11. Align the edge of the brace on the spreader on the same side of the centerline as the center support.
12. Drill two pilot holes through the spreaders. Be careful not to drill into the brace.
13. Square the bench legs to the bench frame and fasten the spreader to the brace with 2 ½″ deck screws. See Figure 85.
14. Repeat steps 11–13 for the other brace.
15. Turn the assembly right side up on a flat surface.
16. Remove the clamps and check for alignment, making sure all of the legs make contact with the flat surface. Adjust the legs if necessary.
17. Maintaining alignment, countersink and drill ⅜″ holes for the carriage bolts, nuts, and washers as described before.
18. Insert the carriage bolts in the holes, place the washers and nuts over the ends and tighten. See Figure 86.

Insert the carriage bolts in the holes...

Figure 85
Figure 85 illustrates step 13

Figure 86
Figure 86 illustrates step 18

Figure 87
Figure 87 illustrates step 1

Attach the Bench Seats to the Bench Frames

The final step in the assembly of the curved picnic table is attaching the bench seats to the bench frames. The concave seats you cut earlier will be attached to the bench frame assemblies with deck screws. Make sure the seats are properly centered.

PROCEDURE

1. Center the concave seat plank lengthwise on the bench frame assembly, with the curved edge of the plank overhanging the legs equallly. See Figure 87.
2. Fasten the concave seat plank in place with five 2 ½″ deck screws, two in each end and one in the center.
3. Center the straight seat plank on the frame ¼″ away from the concave seat plank and fasten in place with six 2 ½″ deck screws.
4. Center the convex seat plank on the frame ¼″ away from the straight seat plank and fasten in place with five 2 ½″ deck screws.
5. Repeat steps 1–4 for the other bench.

Sand and Finish as Desired

Once the entire picnic table has been assembled, you can sand and finish it as you wish. The amount of sanding and the type of finish you select will depend on how you want the picnic table to be used and how you want it to look. Since most picnic tables are used outdoors, you may prefer a relatively rough look. However, others may want their table to have the very smooth, professional look of indoor furniture.

alternate project 2 Folding Picnic Table

Since they can be moved and stored with relative ease, a folding picnic table can be more portable and convenient than a standard picnic table. This folding picnic table is built in two sections that mate together. Both sections have their own seat and table surface. The seat brace on one section is intentionally made smaller than the other so that the two sections can slide together.

This project uses cedar, redwood, or some other exterior grade lumber. Pilot holes should be drilled before installing any of the screws. Drilling the pilot holes helps the screw pass through the first piece and draw the second piece to it. It also helps prevent the wood from splitting.

You will need the following materials:

- (2) 2 x 6 x 12′-0″ cedar or other exterior grade lumber
- (1) 2 x 6 x 10′-0″ cedar or other exterior grade lumber
- (3) 2 x 6 x 8′-0″ cedar or other exterior grade lumber
- (3) 2 x 4 x 8′-0″ cedar or other exterior grade lumber
- (3) 2 x 4 x 12′-0″ cedar or other exterior grade lumber
- (4) 3/8″ x 6″ dowel pegs
- Wood glue
- Damp rag
- (2) 8″ x 1 1/2″ weather resistant strap hinges with screws
- (8) 3/8″ x 3 1/2″ galvanized carriage bolts with nuts and washers
- (200) 2 1/2″ deck screws
- (2) 3/8″ x 6″ wood dowels

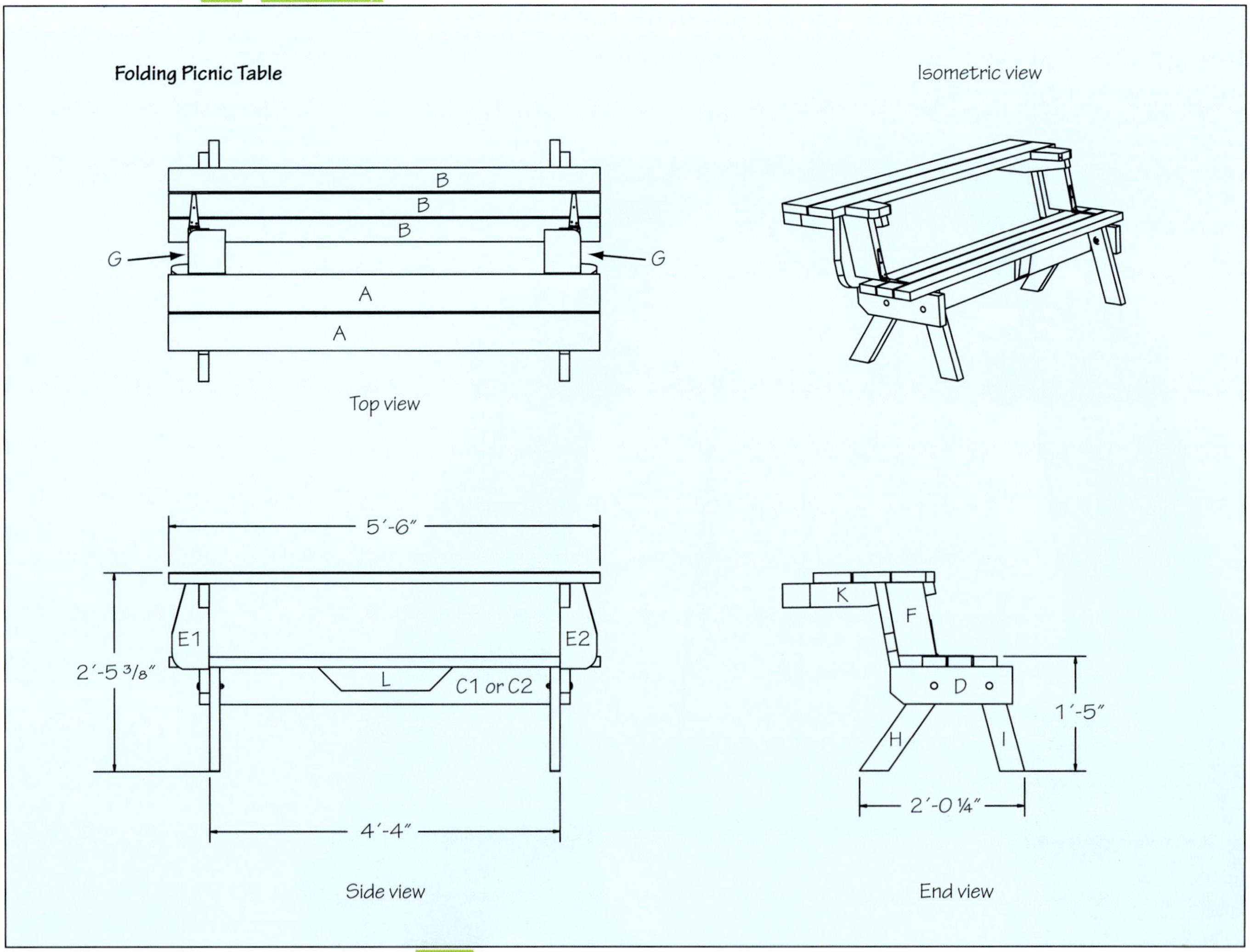

Folding Picnic Table

Isometric view

B
B
B
G
G
A
A

Top view

5′-6″
E1
E2
2′-5 3/8″
L
C1 or C2
4′-4″

Side view

K
F
D
1′-5″
H
I
2′-0 1/4″

End view

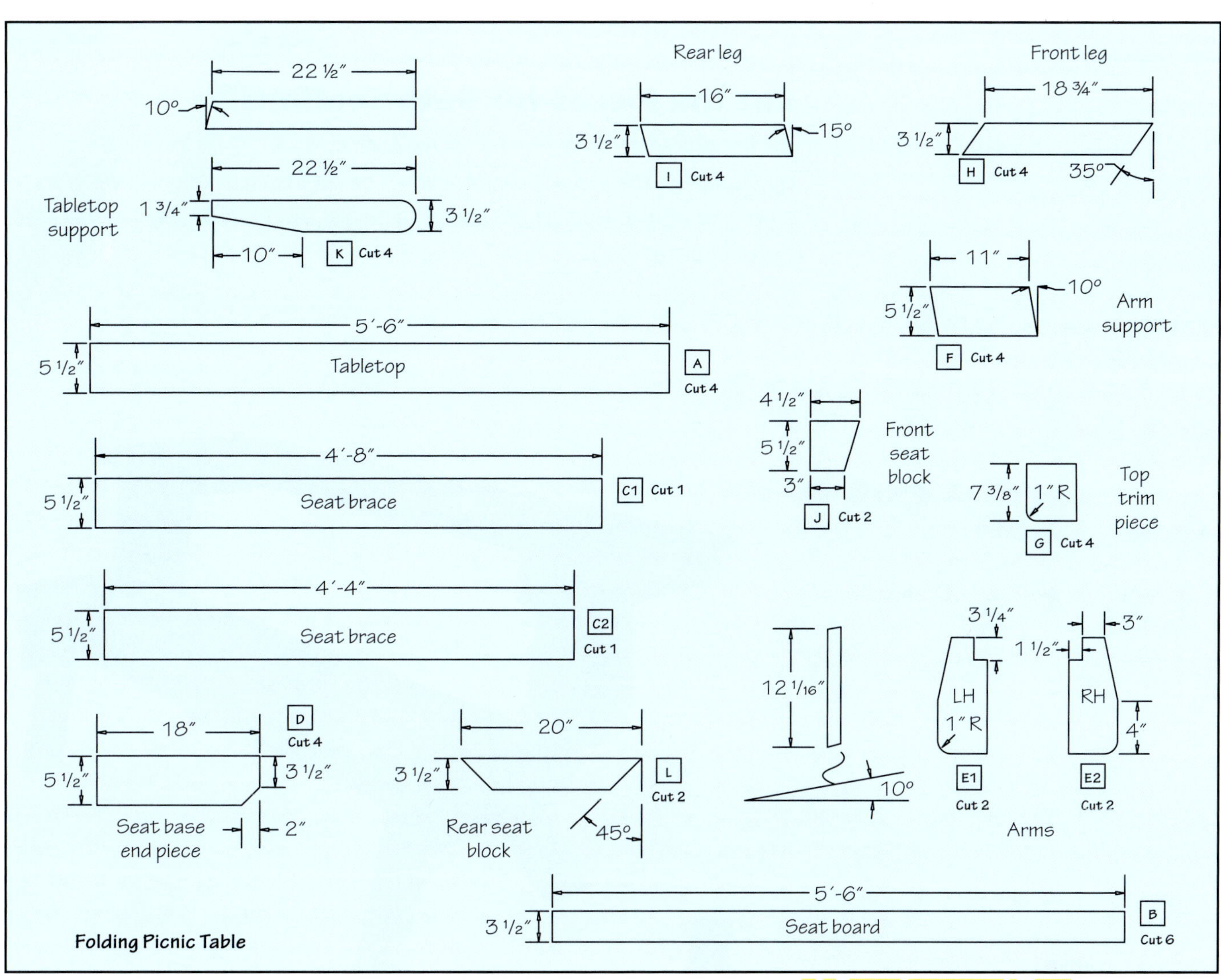

You will need the following tools:

- Adjustable wrench
- Circular saw or miter saw
- Clamps
- Combination square
- Drill index
- Electric drill
- Pencil
- Tape measure
- Sanding block
- Sandpaper, 120 grit and 220 grit
- (2) Sawhorses
- Screw gun
- Sliding T-bevel
- Speed square
- Wood rasp

Cut the Pieces for Both Sections

The first procedure in construction of the folding picnic table is to measure and cut the various pieces. Each component will be cut from a length of cedar or some other exterior grade lumber using either a circular saw or miter saw. You will need to cut 44 pieces in all.

Figure 88
Figure 88 illustrates step 1

Assemble the Seat Frame for the First Section

To assemble the first section of the folding table you will start by marking the end pieces for the seat frame. These pieces are then attached to the seat braces using deck screws. It is necessary to drill pilot holes before inserting the screws.

PROCEDURE

1. On two 2 x 6 x 18″ end pieces for the seat base assembly, measure and mark a 10 ½″ line across the inside face. Measure this line from the angled end of the piece. See Figure 88.
2. Draw another line, parallel to the line drawn in step 1, at 12″. See Figure 89.

Figure 89
Figure 89 illustrates step 2

Figure 90
Figure 90 illustrates step 3

3. Drill 3 pilot holes in both pieces, centered between the two lines. See Figure 90.
4. Attach the two end pieces to the 2 x 6 x 4′-8″ seat brace with six deck screws through the pilot holes in the end pieces into the seat brace. See Figure 91.

Figure 91
Figure 91 illustrates step 4

Assemble the Seat for the First Section

Once the seat base assembly is complete, the seat for the first section can be assembled. The seat boards will be attached to the end pieces of the seat base assembly with deck screws inserted through pilot holes. Then the seat blocks and legs will be attached to the assembly and the legs fastened to the seat base end pieces using carriage bolts.

Figure 92
Figure 92 illustrates step 1

PROCEDURE

1. Set the seat frame on the work surface with the beveled corners down. See Figure 92.
2. Measure and make a mark 5 ½″ from the front beveled end of both 2 x 6 x 18″ seat base end pieces. See Figure 93.
3. Place the first of the three 2 x 4 x 5′-6″ seat boards up against the line toward the back of the bench. Check for equal overhang on each side.
4. Drill two pilot holes through the seat board over each end piece.

Figure 93
Figure 93 illustrates step 2

Figure 94
Figure 94 illustrates step 5

5. Fasten the seat board to the end pieces through the pilot holes with two deck screws. See Figure 94.
6. Lay out the other two 2 x 4 x 5′-6″ seat boards with a ⅛″ gap between them and the first board working toward the back. Center the boards so that they overlap the end pieces equally. See Figure 95.
7. Repeat steps 4 and 5 to fasten the seat boards to the seat frame end pieces.

Figure 95
Figure 95 illustrates step 6

Figure 96
Figure 96 illustrates step 8

8. Turn the seat assembly over on the work surface. See Figure 96.
9. Center and attach 2 x 6 x 4 ½" front seat block to the front of the seat base brace with two 2 ½" deck screws. See Figure 97.

Figure 97
Figure 97 illustrates step 9

Front seat block

3"

Figure 98
Figure 98 illustrates step 10

10. Center and attach the 2 x 4 x 1′-8″ rear seat block to the rear of the seat base brace with three 2 ½″ deck screws. See Figure 98.
11. Clamp the 2 x 4 x 18 ¾″ front legs and the 2 x 4 x 1′-4″ rear legs to the seat assembly. Position the legs against the seat brace end pieces and the seat brace. See Figure 99.
12. Fasten the legs with four 2 ½″ deck screws through each leg into the end pieces.
13. Drill and bolt each leg to the base end piece with one ⅜″ x 3 ½″ carriage bolt. See Figure 100.
14. Remove the clamps and set the seat assembly aside.

Figure 99
Figure 99 illustrates step 11

Figure 100
Figure 100 illustrates step 13

Assemble the Tabletop for the First Section

To assemble the tabletop you will use a flat work surface to support the pieces. The arm supports will be attached to the tabletop support with deck screws. The completed tabletop assembly is then attached to the seat assembly, with strap hinges.

Figure 101
Figure 101 illustrates step 1

PROCEDURE

1. Place one 2 x 4 x 22 ½″ tabletop support, shown in Figure 101, with the tapered edge facing up on the work surface.
2. Mark the location of one 2 x 6 x 11″ arm support by measuring 1 ⅞″ from the tapered end of the tabletop support. See Figure 102.
3. Attach the arm support flush to the back edge of the tabletop support with three deck screws.

Figure 102
Figure 102 illustrates step 2

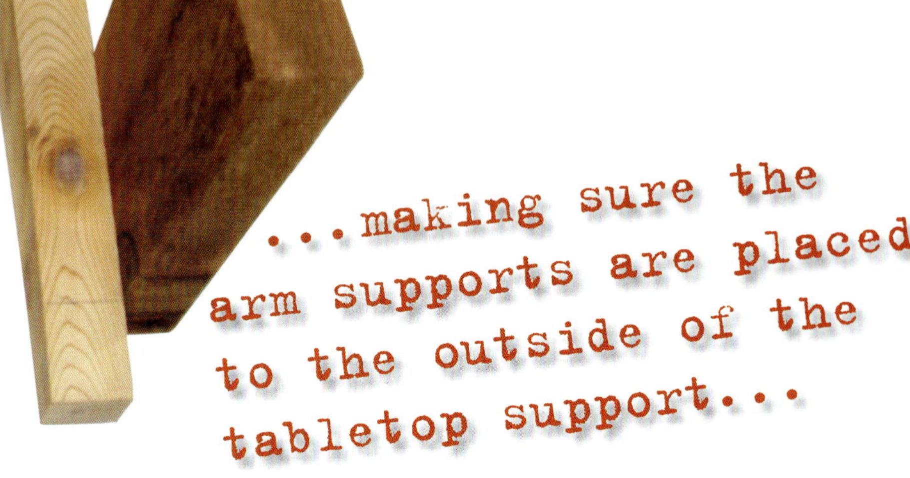

Figure 103
Figure 103 illustrates step 4

4. Repeat steps 1–3 for the other arm support, making sure the arm supports are placed to the outside of the tabletop support, as shown in Figure 103.
5. Set the tabletop support assemblies, shown in Figure 104, over the seat base assembly, aligning the tabletop support assemblies with end pieces for the seat base assembly.

Figure 104
Tabletop support assemblies

6. Measure 7 ½″ back from the tapered end of each tabletop support and make a mark. See Figure 105.
7. Place the first 2 x 6 x 5′-6″ tabletop board with the bottom on the mark away from the tapered end. Check for equal overhang at the ends of the boards. See Figure 106.
8. Fasten the tabletop board to both tabletop supports with two deck screws on each end.
9. Lay out the other 2 x 6 x 5′-6″ tabletop board with a ⅛″ gap and flush with the ends of the first board.

Figure 105
Figure 105 illustrates step 6

Figure 106
Figure 106 illustrates step 7

Figure 107
Figure 107 illustrates step 10

10. Fasten the tabletop boards to the tabletop support pieces with two deck screws at each end. See Figure 107.
11. To locate the pivot point of the strap hinges, rest the arm supports on the bench top with the back edge of the tabletop support flush with the back of the seat base end piece and place a mark at the front of the arm supports. See Figure 108.

Figure 108
Figure 108 illustrates step 11

12. Place a temporary brace under the tabletop as support and install the hinges in the locations determined in step 11. See Figure 109.

Figure 109
Figure 109 illustrates step 12

Figure 110
Figure 110 illustrates step 13

13. Fasten the 2 x 6 x 7 3⁄8″ trim pieces to the back of the arm support with three deck screws, aligning the outside edge with the end of the tabletop boards. See Figure 110. Make sure the spacing between the boards is maintained.
14. Check the length of the arms to make sure the length is correct when in the tabletop position. Fasten the arms to the arm supports as shown on the drawing. See Figure 111.

Figure 111
Figure 111 illustrates step 14

Assemble the Second Section

Assembling the second section of the folding picnic table is essentially a repeat of the previous procedures. However, the seat base brace is 4′-4″ long on the second section instead of 4′-8″ as it is on the first section. After the second side is assembled, the two sections are aligned with dowels.

Assemble the Two Sections

Now that the two separate sections of the folding picnic table are complete, they can be fit together to form the finished table. You will be using dowel pegs and wood glue as fasteners. Once assembly is complete, you will smooth all rough edges with a rasp and a sanding block.

Figure 112
Figure 112 illustrates step 1

PROCEDURE

1. Scribe a line along the center of the edge of the top horizontal 2 x 6 on both tabletop sections. See Figure 112.
2. On the second unit with the 4′-4″ seat brace, mark the centerline of the two tabletop supports on the edge of the 2 x 6 tabletop board intersecting with the line described in step 1. See Figure 113.
3. Drill two holes, approximately 1 ½″ deep, for the dowel pegs at the intersection created in step 2.
4. Glue and insert the dowel pegs into the holes drilled in step 3. See Figure 114.

Figure 113
Figure 113 illustrates step 2

Figure 114
Figure 114 illustrates step 4

5. Mate the two tabletop sections together confirming the alignment at the ends of the tabletop.
6. Mark where the dowels contact the opposing 2 x 6 tabletop board along the centerline drawn in step 1. See Figure 115.
7. Drill two holes, approximately 1 ½″ deep, for the dowel pegs located in step 6. See Figure 116.
8. Smooth all rough edges with a rasp and a sanding block.

Figure 115
Figure 115 illustrates step 6

Figure 116
Figure 116 illustrates step 7

Sand and Finish as Desired

Once the entire picnic table has been assembled, you can sand and finish it as you wish. The amount of sanding and the type of finish you select will depend on how you want the picnic table to be used and how you want it to look. Since most picnic tables are used outdoors, you may prefer a relatively rough look. However, others may want their table to have the very smooth, professional look of indoor furniture.

alternate project 3
Octagonal Picnic Table

Long ago carpenters learned that they could simulate a circular structure or a round piece of furniture by using an octagonal design. While building an octagonal object did not present the complications of working with curved surfaces, the finished product would offer all the same advantages of a round one. This octagonal picnic table applies this principle, and in the process provides practice cutting and assembling accurate angles. The building of this octagonal picnic table requires eight individual processes, including cutting pieces for the table, assembling the tabletop frame, assembling the seat frame, assembling the tabletop, assembling the seat tops, assembling the seat legs, connecting the tabletop, and finishing the table.

You will need the following materials:

- (2) 2 x 6 x 8′-0″ cedar or other exterior grade lumber
- (10) 2 x 6 x 10′-0″ cedar or other exterior grade lumber
- (9) 2 x 4 x 8′-0″ cedar or other exterior grade lumber
- (1) 2 x 4 x 10′-0″ cedar or other exterior grade lumber
- (1) 2 x 2 x 8′-0″ cedar or other exterior grade lumber
- (8) 5⁄16″ x 4 ½″ galvanized carriage bolts with nuts and washers
- (2lb) 2 ½″ deck screws
- (1lb) 3″ deck screws

Note: Pilot holes should be drilled before inserting all screws. While treated lumber could be used, some people may have a reaction to the chemicals in the lumber. Cedar or other exterior grade lumber is less likely to cause an adverse reaction.

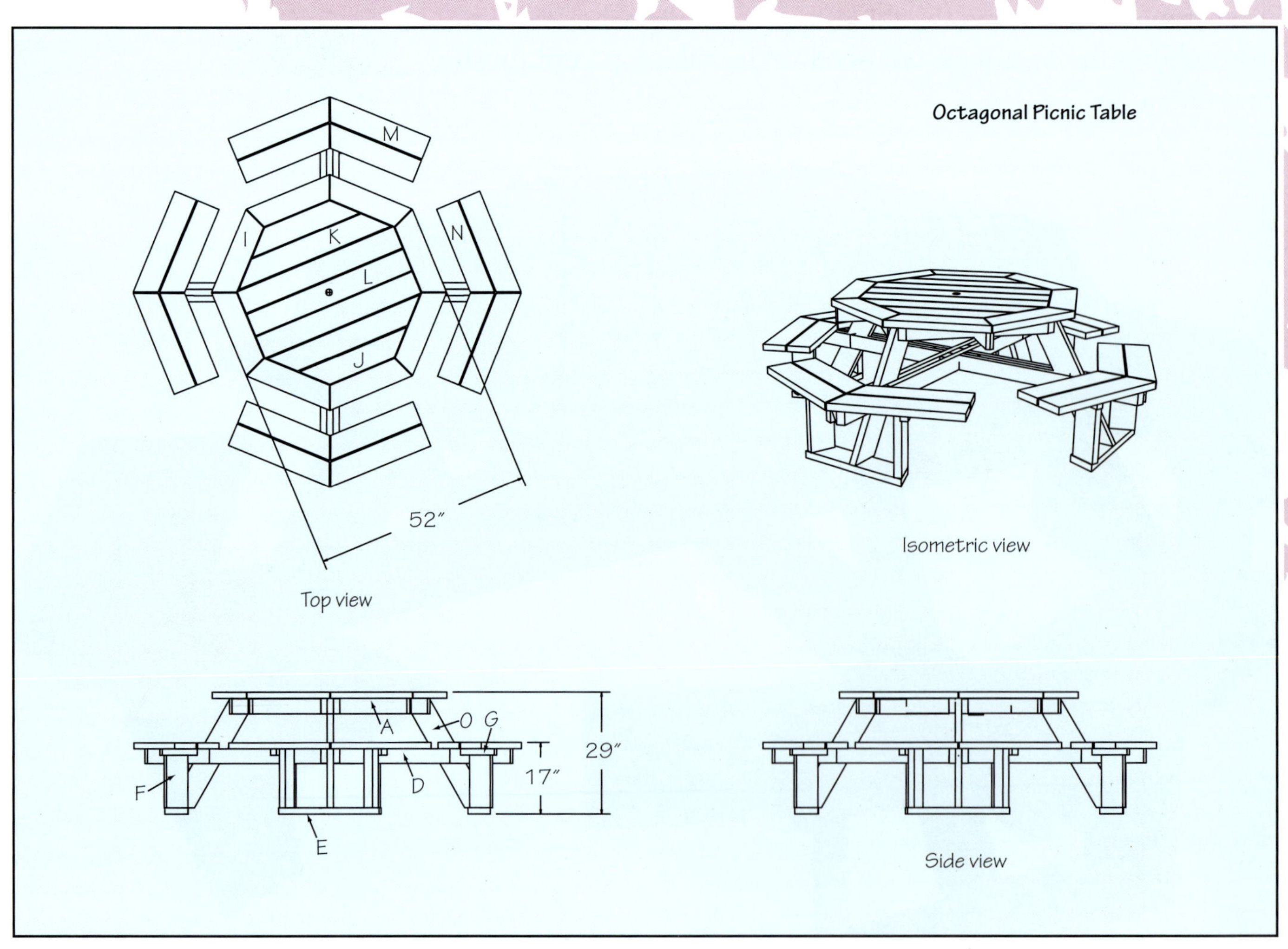

You will need the following tools:

- Band clamp (optional)
- 1″ chisel
- 6″ C-clamps
- Circular saw or miter saw
- Combination square
- Countersink bit for deck screws
- Drill index
- Electric drill
- Framing square
- Mallet
- Pencil
- Tape measure
- Sanding block
- Sandpaper 120 grit and 220 grit
- (2) Sawhorses
- Screw gun
- Sliding T-bevel
- Socket wrench and sockets
- 5⁄16″ spade bit
- Speed Square
- Wood rasp

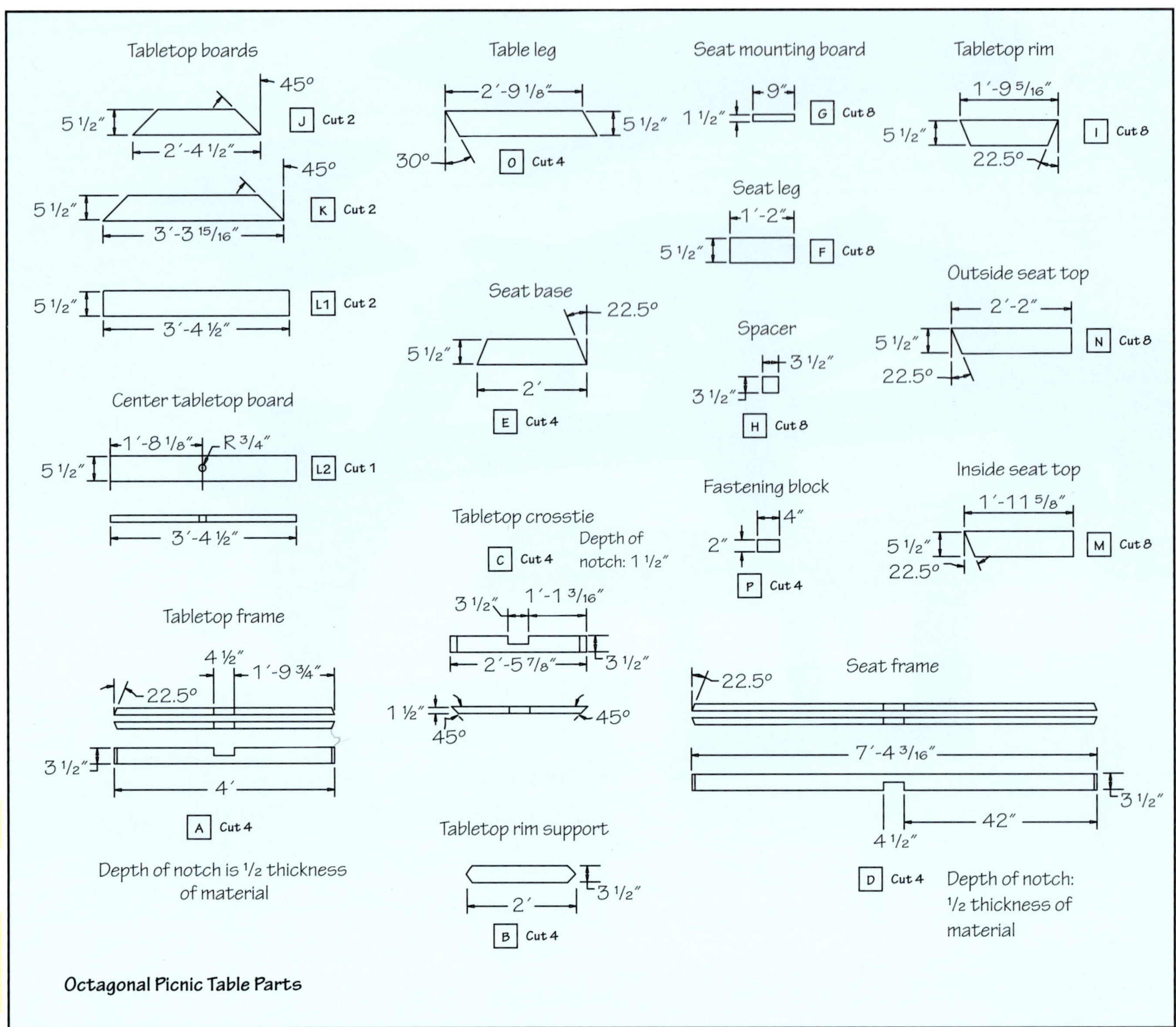

Octagonal Picnic Table Parts

Cut Pieces for the Table

The octagonal table will require 83 separate parts made of cedar or some other exterior grade lumber. You will cut these with a circular saw or miter saw from various sizes and lengths of high quality cedar lumber. When selecting the lumber for this project and cutting the pieces, imperfections should be cut off if possible or the piece should be used with the imperfections facing in a direction that makes them less noticeable. If the lumber has a bow in it, it can be used for the smaller pieces. If it is too bowed, however, it should not be used at all.

Figure 117
Figure 117 illustrates step 1

Assemble the Tabletop Frame

Assembly of the table begins with the tabletop frame. The frame will be constructed with four pieces of 2 x 4 x 4′-0″ beveled and notched lumber. Make sure the bevels face out and the joints are precisely 90°. All pieces including rim supports and crossties will be fastened with corrosion resistant deck screws.

PROCEDURE

1. On a flat and level surface, orient two of the 2 x 4 x 4′-0″ tabletop frame pieces so the notch for the overlap joint faces up. See Figure 117.
2. Check the bevels on both ends of both pieces and make sure they face out.
3. Insert the remaining two 2 x 4 x 4′-0″ pieces of the tabletop frame into the overlap joint. See Figure 118.
4. Check the bevels on both ends of both pieces and make sure they face out.
5. Insert one 2 x 4 x 3 ½″ spacer on each side of the joint for a total of four spacers. See Figure 119.

Figure 118
Figure 118 illustrates step 3

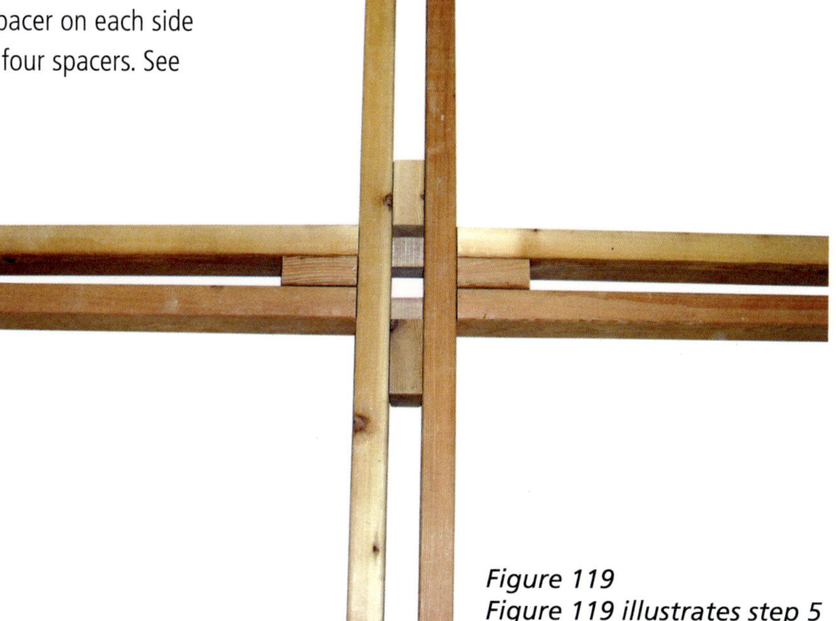

Figure 119
Figure 119 illustrates step 5

6. Verify with a framing square that the angles where the tabletop frame pieces join are 90°. See Figure 120.
7. Fasten the tabletop frame pieces together at the joint with 3″ deck screws through the top of the overlap joint. See Figure 121.

Figure 120
Figure 120 illustrates step 6

Figure 121
Figure 121 illustrates step 7

Figure 122
Figure 122 illustrates step 8

8. Fasten the tabletop supports to the spacers with two 2 ½″ deck screws in each spacer, one from each side. See Figure 122.
9. Install crosstie with notches of dado facing down and fasten as pictured with two 2 ½″ deck screws at each end. See Figure 123.

Install crosstie as pictured...

Figure 123
Figure 123 illustrates step 9

Figure 124
Figure 124 illustrates step 10

10. Install rim support through the dado notch of the crosstie and fasten with 2 ½″ deck screws. See Figure 124.
11. Install 2 x 2 x 4″ fastening blocks in the corners of the tabletop frame support with two 2 ½″ deck screws and to the tabletop rim support with one 2 ½″ deck screw. See Figure 125.
12. Set the tabletop frame aside.

Figure 125
(a) Figure 125 illustrates step 11 for the corners

(b) Figure 125 illustrates step 11 for the rim support

Assemble the Seat Frame

When assembling the seat frame, make sure you place the beveled pieces so the bevels are facing out. Use a framing square to ensure the seat frame joints are precisely 90°. All pieces, including supports and spacers, will be fastened with deck screws.

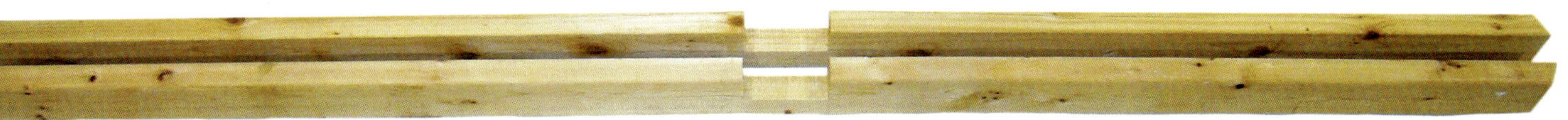

Figure 126
Figure 126 illustrates step 1

PROCEDURE

1. On a flat and level surface, orient two of the 2 x 4 x 7′-4 3⁄16″ seat frame pieces so the notch for the overlap joint faces up. See Figure 126.
2. Check the bevels on both ends of both pieces and make sure they face out.
3. Insert the remaining two 2 x 4 x 7′-4 3⁄16″ pieces of the seat frame into the overlap joint. See Figure 127.
4. Check the bevels on both ends of both pieces and make sure they face out.
5. Insert one 2 x 4 x 3 ½″ spacer on each side of the joint for a total of four spacers.

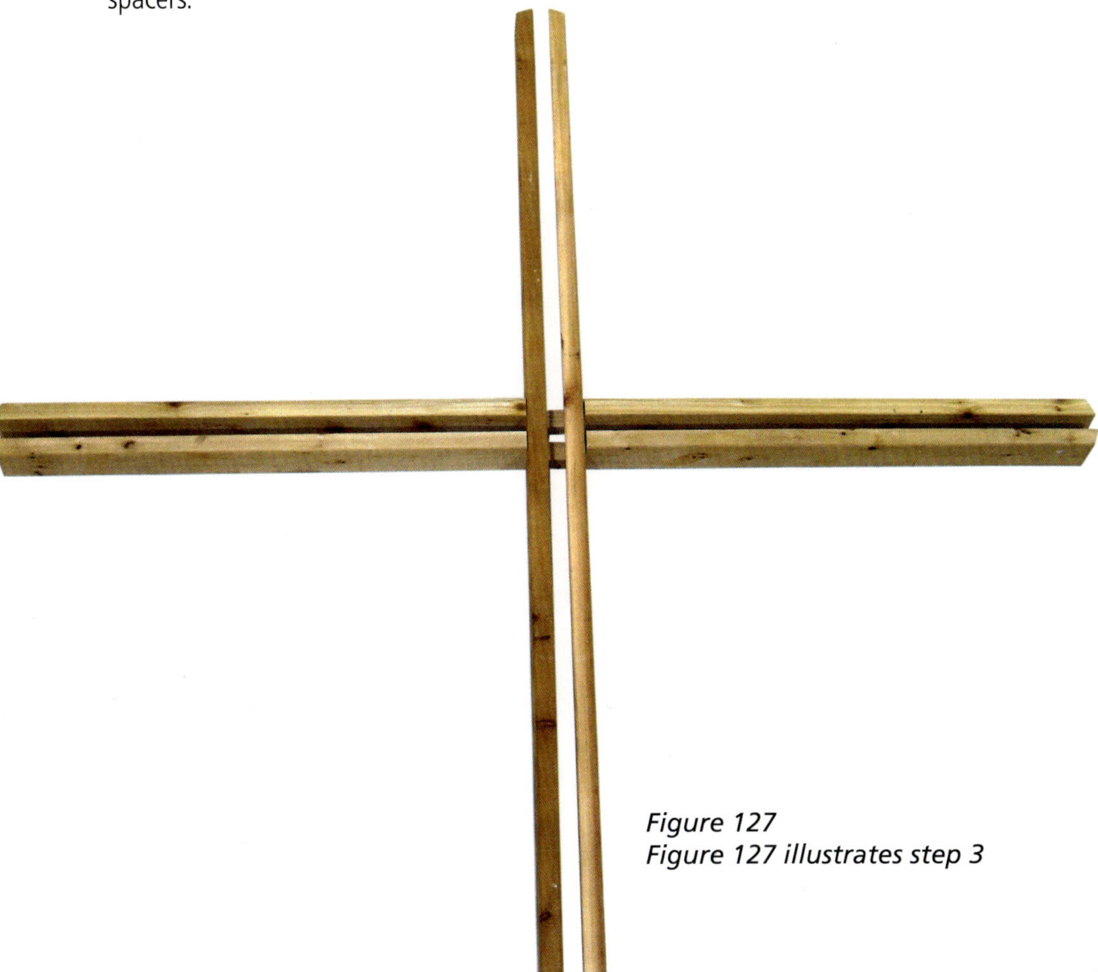

Figure 127
Figure 127 illustrates step 3

6. Verify with a framing square that the angles where the seat frame pieces join are 90°. See Figure 128.
7. Fasten the seat frame pieces together at the joint with four 3″ deck screws through the top of the overlap joint.
8. Fasten the seat frame supports to the spacers with two 2 ½″ deck screws in each spacer, one from each side. See Figure 129.
9. Set the seat frame assembly aside.

Figure 128
Figure 128 illustrates step 6

Figure 129
Figure 129 illustrates step 8

Assemble the Tabletop

Now it is time to assemble the tabletop. First, the rim pieces will be fit together using a template as a pattern. The rim pieces will be fastened with deck screws, which you will countersink so that they are flush with the surface. Once the rim and center boards are in place, the tabletop will be attached to the tabletop frame.

PROCEDURE

1. Lay out the eight 2 x 6 x 1′-9 5⁄16″ tabletop rim pieces. Make sure that the best surface is laid downward.
2. Fit the tabletop rim pieces tightly against the 2 x 4 stops in the template. See Figure 130.
3. Fasten the tabletop rim pieces together with 3″ deck screws driven and countersunk at an angle at each joint.
4. Place the interior tabletop boards within the rim pieces with equal spacing between the boards. Adjust the length of the angled boards as needed to achieve equal spacing. See Figure 131.

Figure 130
Figure 130 illustrates step 2

Figure 131
Figure 131 illustrates step 4

Figure 132
Figure 132 illustrates step 5

5. Center the tabletop frame over the assembled tabletop. See Figure 132.
6. Check the spacing of the tabletop boards to make sure they are spaced uniformly.
7. Fasten the tabletop frame to the tabletop boards by countersinking 1 ¼″ and driving one or two 3″ deck screws at each joint as needed. A spacer block can be temporarily fastened between the tabletop frame ends to keep the assembly centered on the tabletop while fastening the tabletop frame to the tabletop. See Figure 133.
8. Fasten the tabletop crossties to the tabletop boards with 3″ deck screws as needed.
9. Fasten tabletop rim support with four 2 ½" deck screws at each tabletop rim joint and tabletop pieces as needed. See Figure 134.

Figure 133
Figure 133 illustrates step 7

Figure 134
Figure 134 illustrates step 9

The Seat Legs and Table Legs Assembly

The assembly consists of two 2 x 6 x 1′-2″ seat legs and one 2′-9 ⅛″ table leg attached to the seat base with 3″ deck screws. The screws will be driven through the seat base and into each leg. Once all three legs have been attached, the process is repeated for each of the remaining three seat base and leg assemblies.

Figure 135
Figure 135 illustrates step 1

PROCEDURE

1. Fasten a 2 x 6 x 1′-2″ seat leg to each end of the seat base with three 3″ deck screws at each joint, making sure they are flush with the angle of the seat base. See Figure 135.
2. Using one of the 2 x 6 x 2′-9 ⅛″ table legs, hold it on top of the seat base and against the seat leg and make a mark at the top of the seat leg onto the table leg. See Figure 136. This mark establishes the height of the seat frame on the leg.
3. Center the 2 x 6 x 2′-9 ⅛″ table leg on one 2 x 6 x 2′-0″ seat base member.
4. Align the long point of the table leg with the edge of the seat base and fasten the seat base member to the table leg with three 3″ deck screws, driving them from the bottom of the seat base member into the end of the table leg. See Figure 137.
5. Repeat steps 1–4 for the remaining three seat base and table leg assemblies.

Figure 136
Figure 136 illustrates step 2

Figure 137
Figure 137 illustrates step 4

Figure 138
Figure 138 illustrates step 7

6. Insert all 4 table legs of the seat base/table leg assemblies between the 2 x 4 members of the seat frame 1′-1 11⁄16″ from the end of the seat frame assembly. Use temporary blocks of wood to help support the seat frame.
7. Clamp the table leg assemblies in place, aligning them with the marks made in step 2, and fasten with two 4 ½″ carriage bolts, countersinking for the nut and washer. See Figure 138.
8. Fasten one 2 x 2 x 9″ seat mounting board to the outside of each seat leg, flush with the top of the seat leg, holding a ¾″ projection on the outside edge. A ¾″ gauge block can be used for this. See Figure 139. Fasten with two 2 ½″ deck screws.

Figure 139
Figure 139 illustrates step 8

Figure 140
1″ overhang

Assemble the Seat Tops

The seat tops consist of two 2 x 6 x 1′-11 ⅝″ inside seat tops and two 2 x 6 x 2′-2″ outside seat tops. These are attached to the seat frame and seat base assembly using 3″ deck screws. The same process is repeated for each of the picnic table seats.

PROCEDURE

1. Place two 2 x 6 x 2′-2″ outside seat tops on top of the seat frame/seat base assembly and align the mitered joints with the center line of the seat frame assembly. Make sure you have a 1″ overhang from the outside end of the 2 x 2 seat mounting block. See Figure 140.
2. Fasten the two 2 x 6 x 2′-2″ seat tops to the seat frame and seat base assembly with 3″ deck screws. See Figure 141.
3. Place two 2 x 6 x 1′-11 ⅝″ inside seat tops on top of the seat frame/seat base assembly tight against the 2′-2″ seat tops fastened in step 2.
4. Fasten the two 2 x 6 x 1′-11 ⅝″ seat tops to the seat frame and seat base assembly with 3″ deck screws. See Figure 142.
5. Repeat steps 1–4 for the remaining three seats.

Figure 141
Figure 141 illustrates step 2

Figure 142
Figure 142 illustrates a completed step 4

Figure 143
Figure 143 illustrates step 2

. . . you can sand and finish it as you wish. . .

Connect the Tabletop

The final stage in the assembly of your octagonal picnic table is connecting the tabletop. This can be accomplished by carefully centering the tabletop on the leg assemblies and securing them with 3″ deck screws.

PROCEDURE

1. Center the tabletop on the leg assemblies, measuring around the edge to the legs to make sure it is centered. Clamp the tabletop in position.
2. Fasten with two 3″ screws on both sides of the table frame assembly into the table legs. See Figure 143.

Sand and Finish as Desired

Once the entire picnic table has been assembled, you can sand and finish it as you wish. The amount of sanding and the type of finish you select will depend on how you want the picnic table to be used and how you want it to look. Since most picnic tables are used outdoors, you may prefer a relatively rough look. However, others may want their table to have the very smooth, professional look of indoor furniture.

Student Name:______________________________________ Date: ________________

Curved Picnic Table Project Evaluation

PROCEDURE	CRITERIA AS SPECIFIED BY THE PROJECT	POSSIBLE POINTS	SCORE
Picnic Table	Finished length correct	5	
	Finished height correct	5	
	Curvature correct	5	
	Equal spacing of top boards	5	
	Top boards fastened properly	5	
	Top supports located properly	5	
	Leg assemblies located properly (2)	5	
	Leg assemblies square to top (2)	5	
	Location of leg spreaders correct (2)	5	
	Spread of legs at the base correct (2)	5	
	Tabletop angle braces installed properly (2)	5	
	Carriage bolts installed properly	5	
	Subtotal	*60*	
Picnic Bench Seats (2)	Finished length correct	5	
	Finished height correct	5	
	Curvature correct	5	
	Equal spacing of top boards	5	
	Seat boards fastened properly	5	
	Seat frames assembled properly	5	
	Leg assemblies located properly (4)	10	
	Leg assemblies square to top (4)	10	
	Location of leg spreaders correct (4)	10	
	Spread of legs at the base correct (4)	10	
	Angle braces installed properly (4)	10	
	Carriage bolts installed properly	5	
	Subtotal	*85*	
Overall	Edges rounded	5	
	Edges flush	5	
	Edges rounded per procedure	5	
	Joints tight	5	
	No splits and shiners	5	
	Subtotal	*25*	
General	Tool handling	5	
	Followed direction	5	
	Cleaned up	5	
	Safe work practices	5	
	Subtotal	*20*	
		Student score:	

Total Possible Points = 190

Suggested minimum acceptable score: 133 points or ______

Student's Signature: ____________________________ Teacher's Signature: ____________________________

Student Name:______________________________ Date: ______________

Folding Picnic Table Project Evaluation

PROCEDURE	CRITERIA AS SPECIFIED BY THE PROJECT	POSSIBLE POINTS	SCORE
Cut List	Seat pieces (6)	15	
	Tabletop pieces (4)	10	
	Seat brace (2)	5	
	Seat base end pieces (4)	10	
	Side arm (4)	10	
	Arm support (4)	10	
	Trim pieces (4)	10	
	Front leg (4)	10	
	Rear leg (4)	10	
	Support for tabletop (4)	10	
	Block for under the seat (2)	5	
	Subtotal	*105*	
Folding Picnic Table	Finished top length correct (2)	5	
	Finished seat length correct (2)	5	
	Finished top height correct (2)	5	
	Finished seat height (2)	5	
	Equal spacing of top boards	5	
	Top boards fastened correctly	5	
	Trim pieces fastened correctly	5	
	Tabletop support installed correctly (2)	5	
	Arm supports and arms attached correctly (4)	10	
	Seat boards fastened correctly (2)	5	
	Seat base assembled correctly (2)	5	
	Leg assemblies located correctly (4)	10	
	Leg assemblies square to seat top (4)	10	
	Spread of seat legs at the base equal	5	
	Carriage bolts installed correctly (2)	5	
	Top located correctly on seat (2)	5	
	Strap hinges attached correctly (2)	5	
	Dowels located and inserted correctly (2)	5	
	Table top edges flush when two units are connected	5	
	Subtotal	*110*	
Overall	Joints tight	5	
	Edges flush	5	
	No rough edges	5	
	No excess glue	5	
	No splits and shiners	5	
	Subtotal	*25*	
General	Tool handling	5	
	Followed direction	5	
	Cleaned up	5	
	Safe work practices	5	
	Subtotal	*20*	
		Student score:	

Total Possible Points = 260

Suggested minimum acceptable score: 182 points or ______

Student's Signature: ______________________ Teacher's Signature: ______________________

Student Name: ______________________________ Date: ______________

Octagonal Picnic Table Project Evaluation

PROCEDURE	CRITERIA AS SPECIFIED BY THE PROJECT	POSSIBLE POINTS	SCORE
Cut list	Tabletop frame (4)	10	
	Tabletop rim supports (4)	10	
	Tabletop crossties (4)	10	
	Seat frame (4)	10	
	Seat base (4)	10	
	Seat legs (8)	20	
	Seat mounting boards (8)	20	
	Spacers (8)	20	
	Tabletop rim (8)	20	
	Tabletop 2 x 6 x 2′-4 ½″ (2)	5	
	Tabletop 2 x 6 x 3′-3 5/16″ (2)	5	
	Tabletop 2 x 6 x 3′-4 ½″ (3)	7.5	
	Seat tops 2 x 6 x 23 5/8″ (8)	20	
	Seat tops 2 x 6 x 2′-2″ (8)	20	
	Table legs (4)	10	
	Subtotal	*197.5*	
Picnic Table and Built-In Seats	Tabletop cross dimensions equal*	5	
	Finished height correct	5	
	All sides equal length	5	
	Equal spacing of tabletop boards	5	
	Tabletop boards fastened correctly	5	
	Tabletop frame assembled correctly	5	
	Tabletop rim supports installed correctly (4)	10	
	Tabletop crossties installed correctly (4)	10	
	Seat top correct length (4)	10	
**Because the octagonal table has eight sides, you need to measure between all parallel sides. On the octagonal table you will take 4 measurements to ensure that they are all equal.*	Finished seat height correct (4)	10	
	Seat boards fastened correctly	5	
	Seat base assembled correctly (4)	10	
	Leg assemblies square to seat top (4)	10	
	2x4 spacer blocks installed (4)	10	
	Spread of legs at the base equal	5	
	Carriage bolts installed correctly	5	
	Subtotal	*115*	
Overall	Joints tight	5	
	No rough edges	5	
	Edges flush	5	
	No splits and shiners	5	
	Subtotal	*20*	
General	Tool handling	5	
	Followed direction	5	
	Cleaned up	5	
	Safe work practices	5	
	Subtotal	*20*	

Total Possible Points = 352.5

Student score: ______

Suggested minimum acceptable score: 247 points or ______

Student's Signature: ______________________ Teacher's Signature: ______________________

CHAPTER 5
SKATEBOARD RAMP

CONTENTS

Introduction

This chapter offers you a choice of three very different and very interesting projects. You may build a skateboard ramp that may very well receive heavy use in your neighborhood, an Adirondack chair that would look good in any garden, or a portable folding workbench that would make a handy addition to any shop. By completing one or another of these projects, you will expand your knowledge of carpentry tools, materials, and techniques. Each of these projects will also give you a chance to practice the carpentry skills you have already learned.

What's New?

The projects in this chapter are intended to expand your carpentry knowledge and skills. Completing a skateboard ramp, Adirondack chair, or portable folding workbench will require you to learn about processes, procedures, tools, and techniques that, as yet, may be unfamiliar to you. Among these are the following:

Caulking Gun A caulking gun is a manual or power tool used to apply caulk or adhesive that is supplied in a tube. See Figure 1. When pressure is applied to the handle or trigger of the caulking gun, it extrudes the caulk or adhesive through a nozzle of the tube.

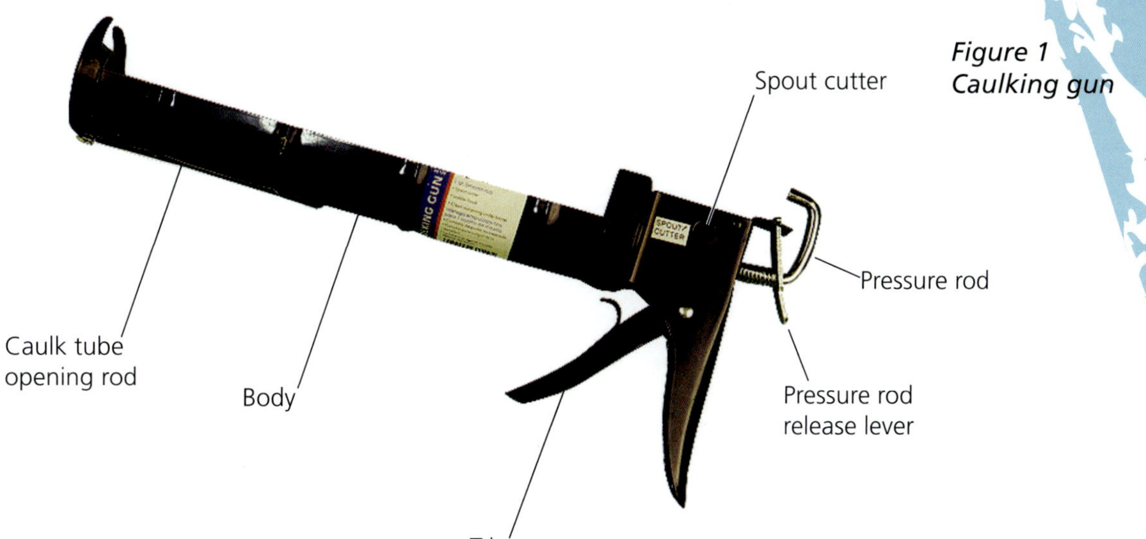

Figure 1
Caulking gun

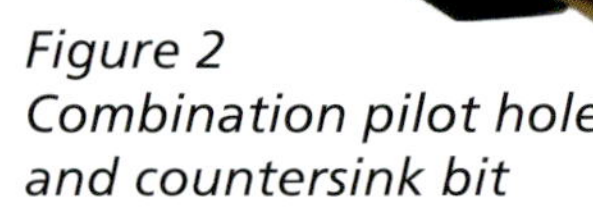

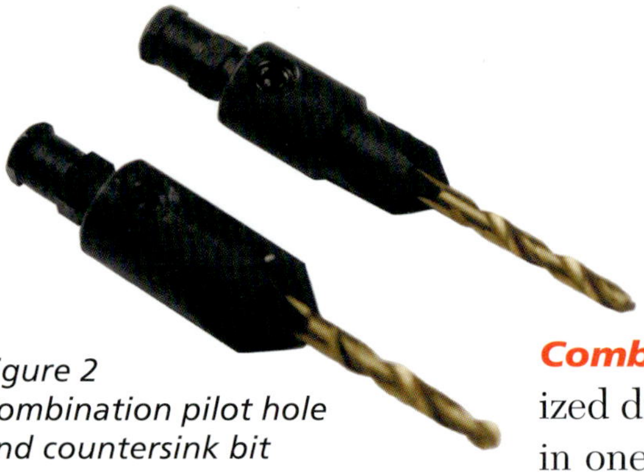

Figure 2
Combination pilot hole and countersink bit

Combination Pilot Hole and Countersink Bit A specialized drill bit for wood that bores a pilot hole and countersinks in one operation is a combination pilot hole and countersink bit. See Figure 2. These bits are sized for specific sized screws. For example, a #8 bit is used for #8 screws.

Figure 3
Exterior wood glue

Exterior Wood Glue Designed for outdoor use, exterior wood glue contains chemicals that help it resist heat, cold, and moisture. Exterior wood glue is an excellent choice of adhesive for constructing outdoor furnishings. See Figure 3.

Panel Adhesive Panel adhesives are specially designed for bonding manufactured panels. When selecting a panel adhesive, refer to the panel manufacturer's specifications. Panel adhesives are usually applied with a caulking gun. See Figure 4.

Figure 4
Panel adhesive

Piano Hinge A piano hinge is a long, narrow hinge that is designed to run the full length of the two surfaces that are to be joined with the hinge. The two leaves of a piano hinge are held together by a pin that allows it to pivot freely. See Figure 5.

Figure 5
Piano hinge

Poplar A type of wood that is relatively soft and light, poplar lumber is easy to cut and work, and it receives paint or stain well. See Figure 6.

Figure 6
Poplar

1 Skateboard Ramp

The curved surface of a skateboard ramp gives roller skaters or skateboarders a place to build up momentum for jumps and other acrobatics. These ramps can provide hours of fun for kids and adults. By following the detailed instructions provided in this section, you can build a highly serviceable and entertaining skateboard ramp of your own, and learn a lot about carpentry in the process. For example, you will learn how to apply panel adhesive with a caulking gun, work with steel pipe and steel plate, and use a countersink bit.

You will need the following materials:

- (1) ¼″ x 4′ x 4′ plywood
- (1) ⅜″ x 4′ x 4′ plywood
- (1) ¾″ x 4′ x 4′ plywood
- (1) ½″ x 2′ x 4′ plywood
- (6) 2 x 4 x 8′-0″ lumber
- (1) 2″ outside diameter galvanized steel pipe 4′ long
- (1) ⅛″ x 10 ⅜″ x 4′ steel plate
- (2) 11-oz tubes of panel adhesive
- (2) 3″ x 3″ x ⅛″ steel 90° brackets
- (24) 1″ deck screws
- (30) 3″ deck screws
- (64) 2″ deck screws

. . . you can build a highly serviceable and entertaining skateboard ramp of your own. . .

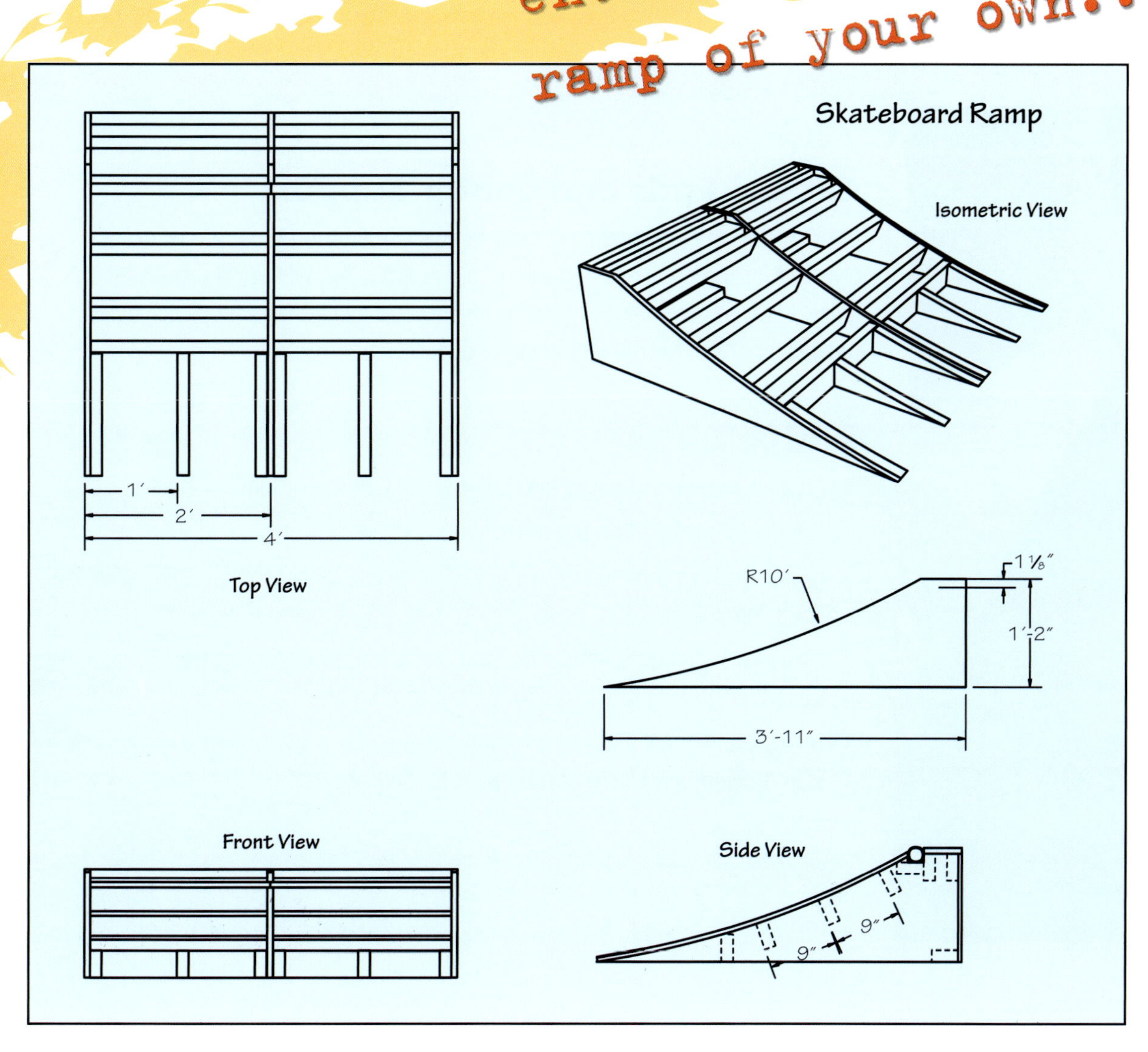

Figure 7
Figure 7 illustrates step 3

You will need the following tools:

- Caulking gun
- Center punch
- Hammer
- Chalk box
- Circular saw
- Clamps
- Combination square
- Countersink bit to cut steel
- Electric drill
- Drill index
- Pencil
- Tape measure
- Saber saw
- (2) Sawhorses
- Screw gun

Figure 8
Figure 8 illustrates step 5

Cut the Ends and Center Support

Construction of the skateboard ramp begins with cutting three pieces of ¾″ plywood for the two ends and the center support. You will use a saber saw to make a curved cut on each of the three plywood pieces. The first piece cut will provide a pattern for the remaining two pieces.

PROCEDURE

1. Prepare a work surface by laying two 2 x 4s on top of two sawhorses.
2. Lay the ¾″ plywood on top of the work surface.
3. Layout and cut three pieces 14″ x 4′-0″ from the ¾″ plywood, cutting with the grain. See Figure 7.
4. Measure 5″ along the 4′-0″ side from one of the pieces cut in step 3 and make a mark at the edge. This will be the top of the ramp.
5. Using the ⅛″ steel plate as a straightedge, scribe a line from the mark made in step 4 diagonally to the opposite corner. See Figure 8.

6. Measure the diagonal line and make a mark at the center.
7. Scribe a line perpendicular to the diagonal line at the center marked in step 6, toward the bottom of the ramp. See Figure 9.
8. Measure and mark 2″ along the perpendicular line. See Figure 10.

Figure 9
Figure 9 illustrates step 7

Figure 10
Figure 10 illustrates step 8

Figure 11
Figure 11 illustrates step 9

Figure 12
Figure 12 illustrates step 10

9. Tack a nail at the mark made in step 8. See Figure 11.
10. Stand the steel plate behind the nail and bend the ends to the 5″ mark made in step 4 and to the bottom corner. This creates the curve in the ramp. See Figure 12.
11. Trace the curve on the plywood on the nail side of the metal plate.
12. Remove the metal plate and nail.
13. Cut the curve with the saber saw. See Figure 13.

Cut the curve with the saber saw.

Figure 13
Figure 13 illustrates step 13

Figure 14
Figure 14 illustrates step 14

14. Locate the notch for the pipe by drawing a square line at the 5″ mark approximately 2″ long. See Figure 14.
15. Measure 1 ⅛″ on the line drawn in step 14 and place a mark. See Figure 15.
16. Draw a square line through the 1 ⅛″ mark back to the curved surface with the combination square. See Figure 16.

Figure 15
Figure 15 illustrates step 15

Figure 16
Figure 16 illustrates step 16

Figure 17
Figure 17 illustrates step 18

17. Cut out the notch for the pipe formed by the lines made in steps 14–16 with a saber saw.
18. Use this piece of plywood as a pattern to mark and cut the other two 14″ x 4′-0″ pieces to match. See Figure 17. These three pieces will serve as the two ends and the center support. Make sure to trace the notch as shown in Figure 18.

Figure 18
Figure 18 illustrates step 18

Cut the Blocking and Wedges

Next you will cut blocking and wedges to support the panels and other parts of the skateboard ramp. Usually, solid blocks of wood are screwed or glued between the panels for use as blocking. In this case, the blocking will consist of 16 separate pieces cut from 2 x 4 lumber. The wedges will support the end of the ramp.

PROCEDURE

1. Using four 2 x 4 x 8′-0″ boards, cut sixteen 2 x 4 blocks 22 ⅞″ long to be used for blocking between the plywood panels.
2. On a 2 x 4 x 8′-0″ board measure and mark a square line 16″ from a square end. See Figure 19.
3. Draw a diagonal line across the board from one corner of the square end to the marked line. See Figure 20.

Mark the wedges.

Figure 19
Figure 19 illustrates step 2

Figure 20
Figure 20 illustrates step 3

Figure 21
Figure 21 illustrates step 4

4. Cut down the center of the diagonal line with the circular saw. See Figure 21.
5. Cut along the squared line made in step 2. See Figure 22.
6. Repeat steps 2–5 two more times to produce six wedges. These will be used to support the end of the ramp.

Figure 22
Figure 22 illustrates step 5

Assemble the Sides and Blocking

Now that the sides and blocking are cut, they can be assembled. This process will require two people. You will need a screw gun as well as both 2″ and 3″ deck screws.

PROCEDURE

1. Stand two of the three plywood support sections upright.
2. Fasten two plywood sections to the first block using the 2″ deck screws and the screw gun. Use two screws through the plywood into each end of the block. See Figures 23a, b.

Figure 23
Figures 23a and 23b illustrate step 2

3. Fasten the third plywood section to the second block using the 2″ deck screws and the screw gun. Use two screws through the plywood into one end of the block. Angle screw through the center plywood section into the second block. See Figure 24.

Figure 24
Figure 24 illustrates step 3

4. Measure up the curve 16″ from the bottom end of the ramp and place a mark on each end panel. See Figure 25.
5. Use a chalk line to snap a line through the marks on each end panel, marking the center panel.
6. Attach one 16″ tapered wedge to the face of a 22 ⅞″ block at each end using two 3″ deck screws for each wedge.
7. Center and fasten the third wedge between the wedges fastened in step 6 with two 3″ deck screws. See Figure 26.
8. Repeat steps 6 and 7 for the second wedge assembly.

Figure 25
Figure 25 illustrates step 4

Figure 26
Figure 26 illustrates step 7

Figure 27
Figure 27 illustrates step 9

9. Position one wedge assembly between an end panel and the center panel along the line established in step 5 and fasten in place with 2" screws on each end. See Figure 27.
10. Fasten the other wedge assembly in place through the plywood with 2″ screws on each end. See Figure 28. Attach the blocking to the center panel using 3″ deck screws placed at an angle.
11. Position a 22 ⅞″ block on edge between an end panel and the center panel, aligning the top edge of the block with the top edge of the panels and the face of the block with the vertical edge of the notch for the pipe .
12. Fasten in place with 2″ deck screws through the plywood panels. See Figure 29.

Fasten in place...

Figure 28
Figure 28 illustrates step 10

Figure 29
Figure 29 illustrates step 12

13. Repeat steps 11 and 12 to install the block between the center panel and other end panel.
14. Position two 22 ⅞″ blocks horizontally between the panels, against the blocks installed in steps 11–13, and align the top edges of these blocks with the bottom edges of the cutout.
15. Fasten in place with 2″ deck screws through the panels and 3″ deck screws through the vertical 2 x 4 blocks installed in steps 11–13.
16. Position two 22 ⅞″ blocks horizontally between the panels, against the blocks installed in steps 11–13, and align the top edges of those blocks with the top edge of the panels. Figure 30 shows the completed step.
17. Install the six remaining blocks equidistant between the pipe notch and the 2 x 4s of the wedge assembly with 2″ deck screws. Figure 31 shows step 17 completed.

Figure 30
Figure 30 shows step 16 completed

Figure 31
Figure 31 shows step 17 completed

Cut and Install the Back Supports

The back of the skateboard ramp will consist of ½″ plywood held in place with 2″ deck screws. The plywood back will be attached to 2 x 4 supports fastened to the insides of the end pieces and to one side of the center piece.

Figure 32
Position of 2 x 4 supports

PROCEDURE

1. Cut three 2 x 4 x 11″ supports.
2. Fasten the 2 x 4 x 11″ supports with 2″ deck screws to the insides of each end piece and on one side of the center piece. As shown in Figure 32, the 2 x 4 supports should be vertical, between the blocks, with the face against the plywood.

Cut and Install the Plywood Back

Figure 33
Figure 33 illustrates step 3

PROCEDURE

1. Cut a piece of ½″ plywood 14″ x 4′-0″. Be sure to cut with the grain.
2. Place the plywood against the back of the ramp.
3. Align the edges and fasten the plywood to the 2 x 4 supports and spreader blocks with 2″ deck screws. See Figure 33.

Installing the Top and the Plywood for the Ramp

Now it is time to cut and install the top and the plywood face for the ramp. The top will consist of ¾″ plywood, and the ramp will consist of a 4′-0″ x 4′-0″ piece of ⅜″ plywood and a 4′-0″ x 4′-0″ piece of ¼″ plywood. The pieces will be held in place with both deck screws and adhesive.

Figure 34
Position of plywood

PROCEDURE

1. Turn the assembly upright.
2. Cut a piece of ¾″ plywood 5 ½″ x 4′-0″.
3. Place it on the top flush with each side and against the pipe as shown in Figure 34.
4. Fasten with 2″ deck screws into the 2 x 4 blocking. See Figure 35.

Figure 35
Figure 35 illustrates step 4

5. Apply a bead of the panel adhesive on the 2 x 4 blocks on the curve of the ramp, the wedges, and the tops of the support plywood. Do not apply adhesive to the last 8″ of the wedges at this time. See Figure 36.
6. Center the 4′-0″ x 4′-0″ x ¼″ plywood onto the assembly against the pipe and align it to the edges of the side panels with the grain running across the assembly. See Figure 37.

Figure 36
Figure 36 illustrates step 5

Figure 37
Figure 37 illustrates step 6

7. Screw the ¼″ plywood to the 2 x 4 blocking with 2″ deck screws, starting at the top of the ramp and working to the bottom. Place the screws 2″ from each end and 4″ on center thereafter. Do not screw into the wedges at this time. See Figure 38.
8. Apply construction adhesive over the ¼″ plywood. See Figure 39.

Figure 38
Figure 38 illustrates step 7

Figure 39
Figure 39 illustrates step 8

Figure 40
Figure 40 illustrates step 9

9. Center the 4′-0″ x 4′-0″ x ⅜″ plywood onto the assembly against the pipe and align it to the edges of the side panels with the grain running across the assembly. See Figure 40.
10. Screw the ⅜″ plywood through the ¼″ plywood into the 2 x 4 blocking with 2″ deck screws, starting at the top of the ramp and working to the bottom. Place the screws approximately 1″– 2″ from each end and 4″ on center thereafter. Do not screw into the wedges at this time. See Figure 41.

Figure 41
Figure 41 illustrates step 10

Do not screw into the wedges at this time.

Install the Pipe

To install the pipe, you will need to attach two 90° brackets, one at each end of the pipe. Be sure to avoid drilling through the top of the pipe.

PROCEDURE

1. Lay the assembly upright.
2. Set the pipe in position in the notch.
3. Place one leg of the 90° bracket in the bottom of the pipe and the other leg tight to the plywood.
4. Install two 3" deck screws in the holes of the bracket through the plywood and into the supporting 2 x 4 behind the plywood. See Figure 42.
5. Repeat steps 3 and 4 for the other 90° bracket on the opposite side.

Figure 42
Figure 42 illustrates step 4

Install the Metal Plate

In this final stage of construction you will complete the process of attaching the ramp. You will also install the steel plate to protect the edge of the ramp using screws and adhesive. You will countersink each screw so that its head is flush with the surface.

Figure 43
Figure 43 illustrates step 1

PROCEDURE

1. Measure 3′-6 ½″ from the back edge of the ramp along the bottom and make a mark on the curved plywood ramp on each side. See Figure 43.
2. Snap a line across the curved surface of the ramp through the marks made in step 1. See Figure 44.

Figure 44
Figure 44 shows step 2 completed

Figure 45
Figure 45 illustrates step 4

3. Set the circular saw to cut ⅝″ deep. Be careful to set it accurately so as not to overcut into the wedges.
4. Cut along the line marked in step 2. See Figure 45.
5. Finish attaching the plywood by running 2″ deck screws through the curved plywood ramp into the wedges.
6. Mark the center of each wedge on the surface of the curved plywood ramp.
7. Cut the left over piece of ½″ plywood to 5″ wide.
8. Apply adhesive to the tops of the wedges.
9. Butt the ½″ plywood against the curved plywood ramp and align the edges. See Figure 46.

Figure 46
Figure 46 illustrates step 9

10. Attach the ½″ plywood to the wedges with 1″ deck screws. Be careful to install the deck screws so that they do not go entirely through the wedges. See Figure 47.
11. Lay the steel plate on the ½″ plywood, butting it up against the curved plywood ramp and aligning the edges. See Figure 48.
12. Transfer the wedge center line marks from the plywood across the steel plate.
13. On the centerlines place marks at ½″ and 4″ from where the steel plate meets the plywood.
14. Use a center punch and hammer to dimple the steel plate at the marks.
15. Drill a 3⁄16″ diameter hole in the steel plate at each dimple.
16. Insert a steel countersink into the electric drill and countersink each hole in the steel plate to receive the head of the deck screw.
17. Apply adhesive to the surface of the ½″ plywood.
18. Butt the steel plate against the edge of the curved plywood ramp on top of the ½″ plywood, aligning the edges.
19. Fasten in place with 1″ deck screws.

Figure 47
Figures 47 illustrates step 10

Figure 48
Figure 48 illustrates step 11

alternate project 2 Adirondack Chair

The high-back wooden chairs you sometimes see in gardens or out on the lawns of old, elegant hotels are called Adirondack chairs. They got their name from New York's Adirondack Mountains, where summer resorts once made wide use of them. Building an Adirondack chair will require several advanced techniques, including making and using templates, using a router to round over edges, laying out and cutting angles, and doing complex multi-staged assembly work.

Building an Adirondack chair will require several advanced techniques...

You will need the following materials:

- (1) 1 x 6 x 12′-0″ lumber for back slats
- (1) 2 x 4 x 8′-0″ lumber for front spacer, back spacer, lower back frame, two arm braces
- (5) 1 x 2 x 8′-0″ lumber for seat slats
- (1) 2 x 6 x 8′-0″ lumber for two arms, upper back frame
- (1) 2 x 6 x 12′-0″ for two chair frames, two legs
- (4) 3/8″ x 4″ galvanized carriage bolts with washers and nuts
- (2 lbs) 2″ deck screws
- (2 lbs) 3″ deck screws
- Exterior wood glue

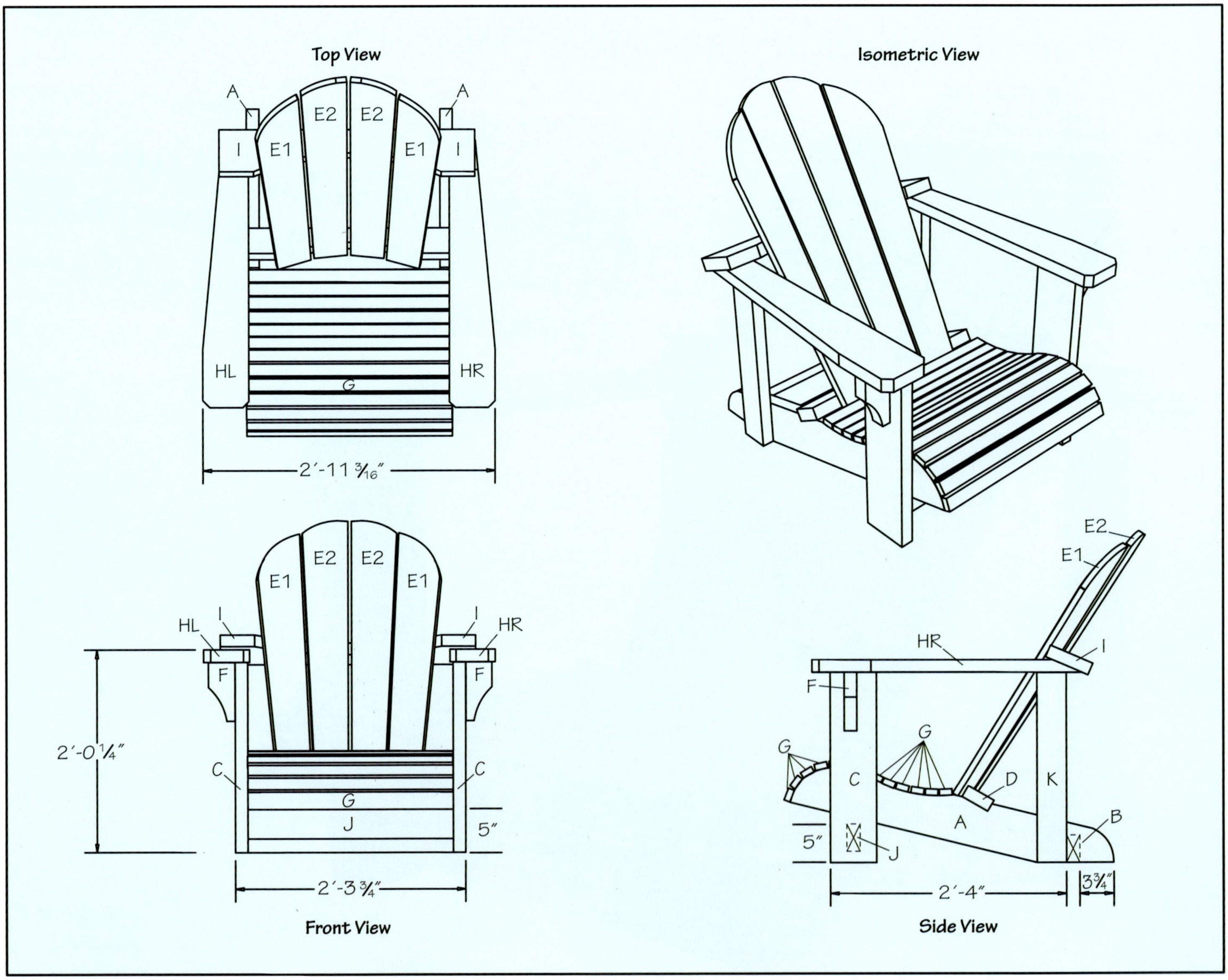

Top View

A A I E1 E2 E2 E1 I HL G HR

2′-11 3/16″

Isometric View

Front View

E1 E2 E2 E1 HL I I HR F F C C G J

2′-0 1/4″ 5″ 2′-3 3/4″

Side View

E2 E1 HR I F G G C D K A B 5″ J 2′-4″ 3 3/4″

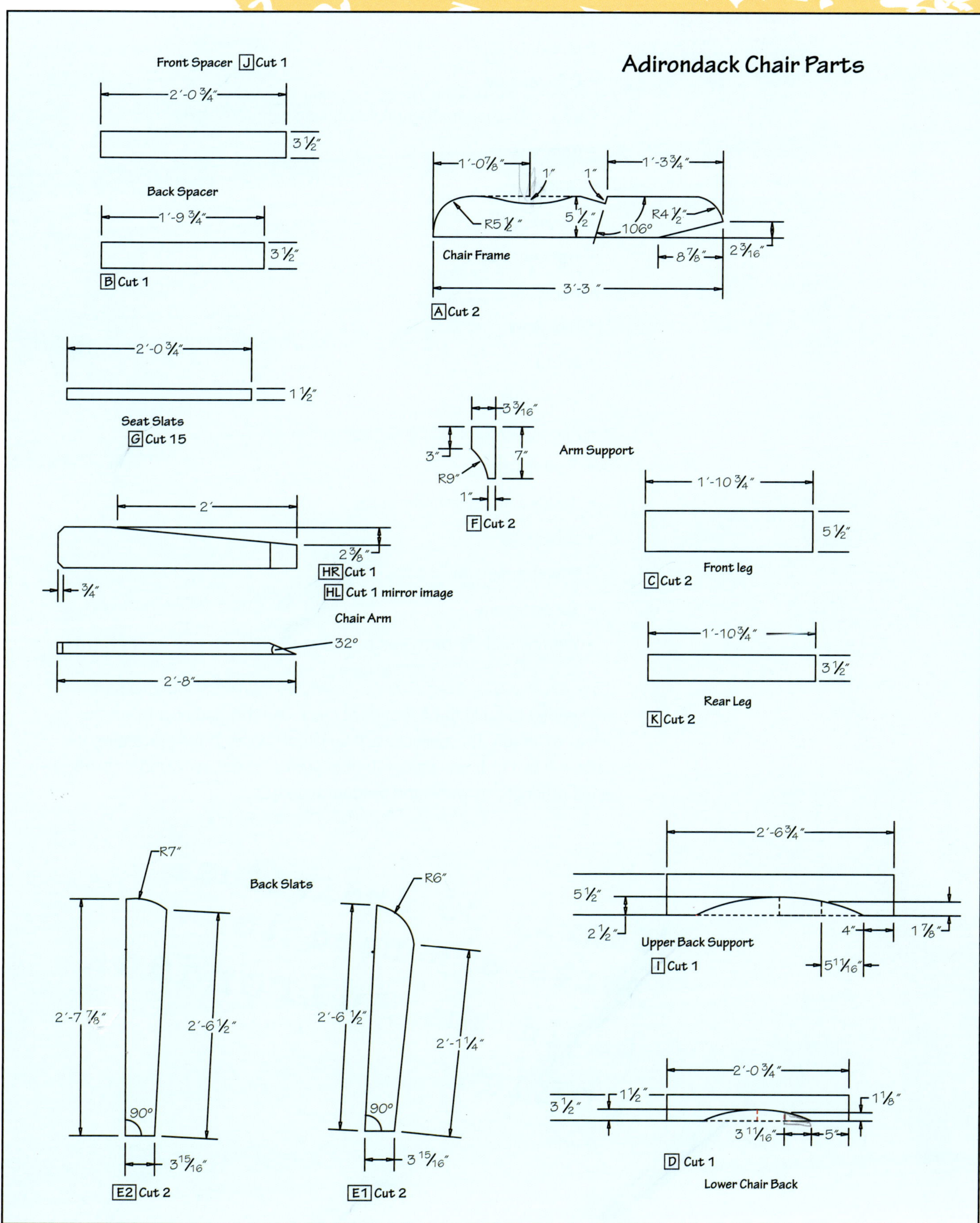

Adirondack Chair Parts
Front Spacer J Cut 1
2'-0 3/4"
3 1/2"
Back Spacer
1'-9 3/4"
3 1/2"
B Cut 1
1'-0 7/8"
1'-3 3/4"
1"
1"
R5 1/2"
5 1/2"
106°
R4 1/2"
Chair Frame
8 7/8"
2 3/16"
3'-3"
A Cut 2
2'-0 3/4"
1 1/2"
Seat Slats
G Cut 15
3 3/16"
3"
7"
R9"
1"
Arm Support
F Cut 2
2'
2 3/8"
3/4"
HR Cut 1
HL Cut 1 mirror image
Chair Arm
32°
2'-8"
1'-10 3/4"
5 1/2"
Front leg
C Cut 2
1'-10 3/4"
3 1/2"
Rear Leg
K Cut 2
R7"
Back Slats
R6"
2'-7 7/8"
2'-6 1/2"
90°
3 15/16"
E2 Cut 2
2'-6 1/2"
2'-1 1/4"
90°
3 15/16"
E1 Cut 2
2'-6 3/4"
5 1/2"
2 1/2"
Upper Back Support
4"
1 7/8"
I Cut 1
5 11/16"
2'-0 3/4"
3 1/2"
1 1/2"
1 1/8"
3 11/16"
5"
D Cut 1
Lower Chair Back

You will need the following tools:

- (2) Adjustable bar clamps
- Circular saw
- #8 x 1 ½″ combination pilot hole and countersink bit
- Compass
- Drill index
- Electric drill
- Framing square
- Hammer
- Handsaw
- Pencil
- Screw gun
- Random orbital sander (optional)
- Tape measure
- Router with ¼″ roundover bit
- Saber saw
- Power miter saw
- Speed Square
- Wrench with ⅜″ open end or socket

Lay Out and Cut Parts Nearly 30 individual wooden parts are needed to build the Adirondack chair. Use this parts list to cut the pieces to length. These pieces will be shaped later in the procedure. The parts will be cut from exterior lumber such as cedar, redwood, or artificial wood products, to withstand outdoor weather.

Nearly 30 individual parts are need to build the Adirondack chair.

Figure 49
Figure 49 illustrates step 2

PROCEDURE

1. Cut the parts to length and label.
2. Lay out and cut one chair frame, part A, to the required shape and dimensions. See Figure 49.
3. Use the chair frame cut in step 2 as a pattern to lay out and cut the second chair frame, part A.
4. Cut the lower chair back, part D, to the required shape and dimensions. See Figure 50.

Figure 50
Figure 50 illustrates step 4

Figure 51
Figure 51 illustrates step 5

Figure 52
Figure 52 illustrates step 6

Figure 53
Figure 53 illustrates step 7

5. Cut the upper chair back, part I, to the required shape and dimensions. See Figure 51.
6. Cut two back slats, part E, to the required shapes and dimensions and use them as templates to make the other two back slats. See Figure 52.
7. Cut one arm, part HL, to the required shape and dimensions. See Figure 53.

Figure 54
Figure 54 illustrates step 8

8. Use the piece cut in step 7 as a pattern to lay out and cut the second arm piece, part H. See Figure 54. Make sure the angles are opposing.
9. Cut one arm support, part F, to the required shape and dimensions. See Figure 55.
10. Use the piece cut in step 9 as a pattern to lay out and cut the second arm support, part F. See Figure 56.

Figure 55
Figure 55 illustrates step 9

Figure 56
Figure 56 illustrates step 10

11. Drill pilot and countersink holes ¾″ from the ends of each seat slat. See Figure 57.
12. Use a router and a ¼″ roundover bit to round over the top edges of the seat slats, all edges of the back slats, the exposed edges of the arm braces, and all edges of both arms. See Figure 58.

Figure 57
Figure 57 illustrates step 11

Figure 58
Figure 58 illustrates step 12

Assemble the Chair

Once cutting and layout are complete, the Adirondack chair can be assembled following the step-by-step procedure outlined below. To make assembly easier, the various parts are identified with a letter. Before assembling the parts review the prints.

Figure 59

Figure 59 illustrates step 2

PROCEDURE

1. Position the chair frames, parts A, upright on a flat surface and space them apart using the lower chair back, part D, as a spreader positioned in the notches.
2. Attach part D to part A with exterior wood glue and four 3″ deck screws, two per chair frame. See Figure 59.
3. Turn the assembly over and position the back spacer, part B, between the chair frames flush with the bottom of the legs and fasten in place with exterior wood glue and four 3″ deck screws, two per chair frame. See Figure 60.

Figure 60

Figure 60 illustrates step 3

4. Position one seat slat, part G, at the front bottom of the chair frames and attach it with exterior wood glue and two 2″ deck screws. See Figure 61.
5. Attach the two legs, part C, to the front spacer, part J, with glue and 3″ deck screws. See Figure 62.
6. Attach the front spacer and leg assembly to the chair frame, part A, with four 3/8″ x 4″ galvanized carriage bolts with washers. See Figure 63.

Figure 61
Figure 61 illustrates step 4

Figure 62
Figure 62 illustrates step 5

Figure 63
Figure 63 illustrates step 6

7. Attach the arm supports, parts F, flush with the top of the legs, part C, and centered left to right with exterior wood glue and three 3″ deck screws each through the leg and into the brace. See Figure 64.
8. Place the upper chair back, part I, on the work surface. Scribe a parallel line 1 3⁄8″ away from the straight edge of the upper chair back along its entire length.
9. Measure and place a mark 3″ from each end of the upper chair back, part I, on the scribed line marked in step 8.
10. Draw a line perpendicular to the straightedge of the upper chair back, part I, through each mark made in step 9. See Figure 65.
11. Apply glue to the upper chair back, part I, where the arm will be attached.

Figure 64
Figure 64 illustrates step 7

Figure 65
Figure 65 illustrates step 10

Figure 66
Figure 66 illustrates step 12

12. Put the beveled end of the arm, part H, on the upper chair back, part I. Attach the arm, part H, to the upper chair back, part I, with one 2″ deck screw, and one 3″ deck screw. Align the long point of the bevel with the 1 ⅜″ line and the straightedge of the arm to the perpendicular line drawn in step 10. See Figure 66.
13. Repeat steps 11 and 12 to attach the second arm, part H.
14. Install the rear leg, part K, to the chair frame. See Figure 67.
15. Apply glue to the top of the arms and arm braces.

Figure 67
Figure 67 illustrates step 14

16. Position the arm assembly on top of the legs and arm braces, with 2″ of the arm assembly extending beyond the front leg. See Figure 68.
17. Secure the arm assembly to the legs and arm supports with 3″ deck screws. Place three screws into the front leg and arm support and two screws into the back leg on each side. See Figure 69.
18. Place a center mark on the upper chair back, part I, and the lower chair back, part D.

Figure 68
Figure 68 illustrates step 16

Place three screws into the front leg and brace and two screws into the back leg.

Figure 69
Figure 69 illustrates step 17

Figure 70
Figure 70 illustrates step 19

Figure 71
Figure 71 illustrates step 20

19. Beginning from the center and working out to the sides, attach the back slats to the upper chair back and lower chair back with exterior glue and two 2″ deck screws in both the lower chair back and upper chair back. Be sure to keep the slats flush to the bottom of the lower back frame. See Figure 70.
20. Install the remaining seat slats, part G, with exterior glue and 2″ deck screws, starting from the seat slat already in place at the front of the chair frame and continuing to the back slats. See Figure 71.
21. Lightly sand any rough surfaces, and finish as desired.

alternate project

3 Portable Folding Workbench

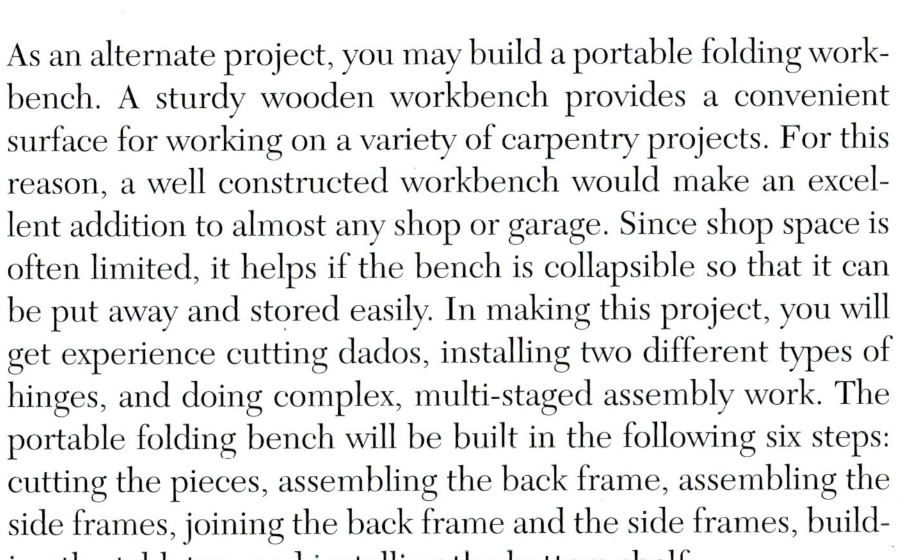

As an alternate project, you may build a portable folding workbench. A sturdy wooden workbench provides a convenient surface for working on a variety of carpentry projects. For this reason, a well constructed workbench would make an excellent addition to almost any shop or garage. Since shop space is often limited, it helps if the bench is collapsible so that it can be put away and stored easily. In making this project, you will get experience cutting dados, installing two different types of hinges, and doing complex, multi-staged assembly work. The portable folding bench will be built in the following six steps: cutting the pieces, assembling the back frame, assembling the side frames, joining the back frame and the side frames, building the tabletop, and installing the bottom shelf.

You will need the following materials:

- (8) 1 x 4 x 8′-0″ poplar or similar lumber
- (1) 1 x 4 x 6′-0″ poplar or similar lumber
- (4) 2 x 4 x 8′-0″ standard or better lumber
- (1) ¾″ x 4′ x 8′ sheet of plywood, finished on one side
- (2) ⅜″ x 1 ½″ dowel pins
- (½ lb) ¾″ construction screws
- (1 lb) 1″ construction screws
- (1 lb) 3″ construction screws
- (1 lb) 6d finish nails
- (6) 3″ strap hinges
- (1) 1″ x 4′ piano hinge with screws
- Exterior wood glue

Note: Pilot holes should be drilled for all of the screws.

Folding Work Bench

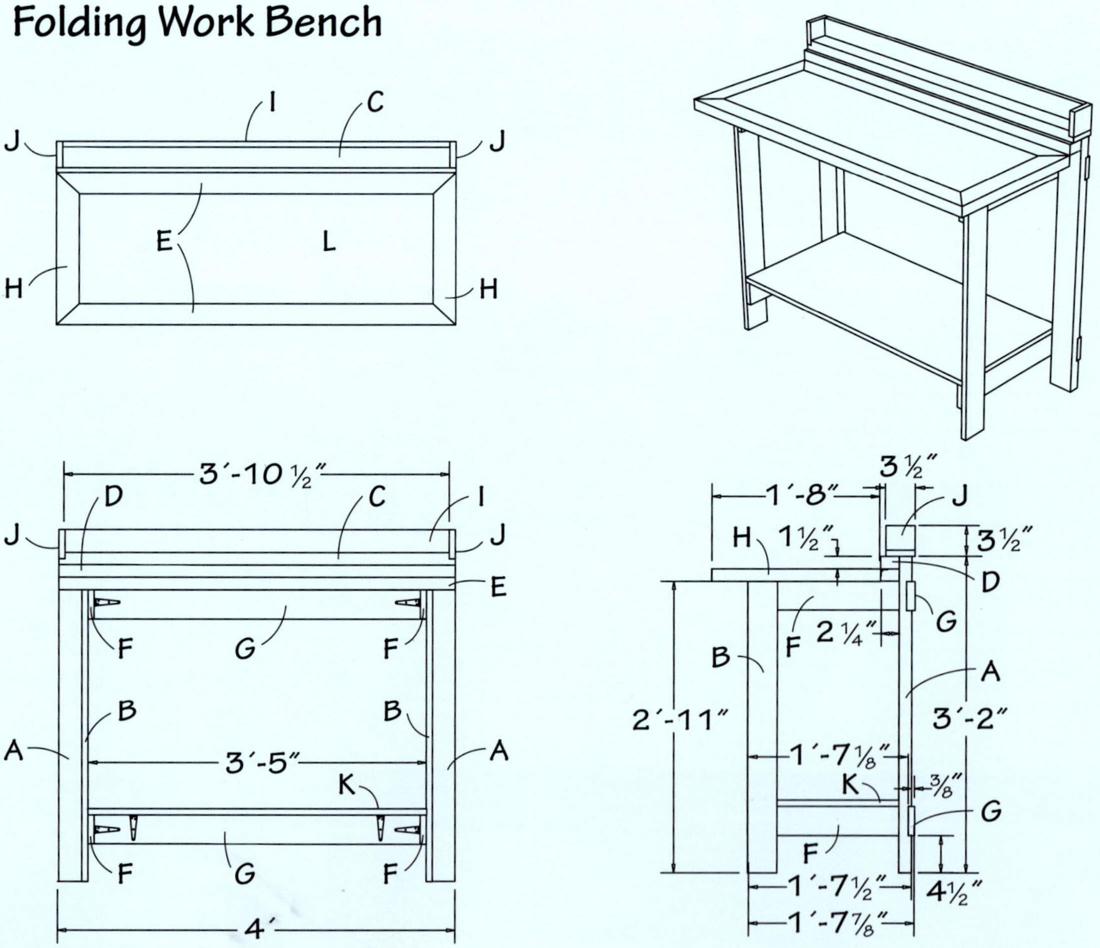

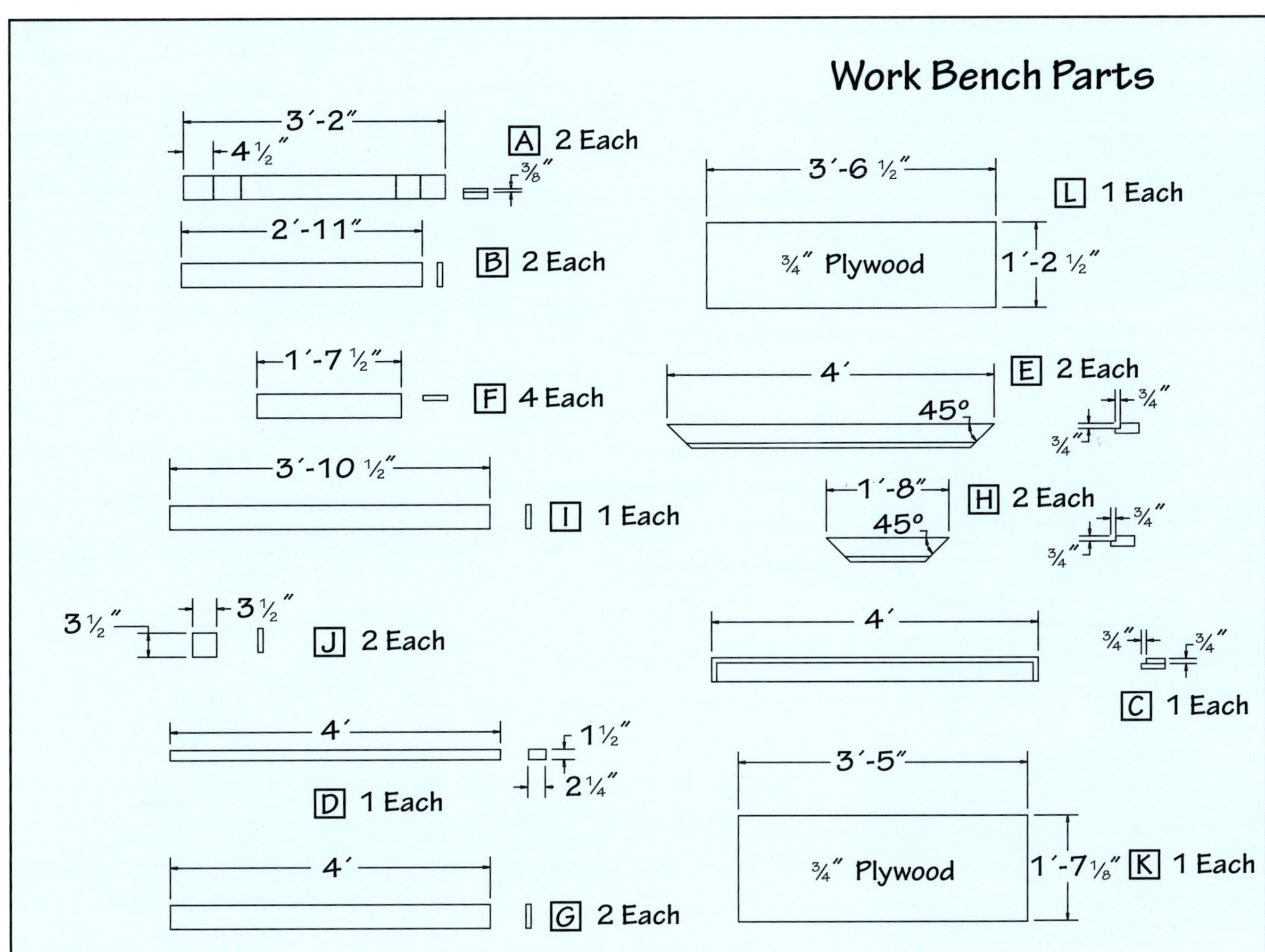

You will need the following tools:

- (2) Bar clamps
- Circular saw
- Combination square
- Drill index
- Electric drill
- Miter saw (optional)
- Pencil
- Tape measure
- Router, ½″ straight bit, ¾″ guide bushing
- Sandpaper
- (2) Sawhorses
- Screw gun
- Wood rasp
- Straightedge
- Adjustable corner clamp

Cut the Pieces

Construction of the portable folding workbench begins with cutting the parts. The workbench requires 21 separate parts cut from poplar or similar lumber and plywood. You will make the cuts using a circular saw or miter saw.

Assemble the Back Frame

The next stage in construction of the portable folding workbench is to assemble the back frame. For this you will use the back legs, back rails, back face, back top, and shelf member pieces you cut in the previous procedure. The various parts will be attached with construction screws.

PROCEDURE

1. Clamp together the two 2 x 4 x 3′-2″ back legs laid flat, side by side, edge to edge with the ends flush.
2. Lay out the top and bottom dados on each of the 2 x 4 x 3′-2″ back legs by measuring from the end of each leg and placing a mark at 4 ½″ and another mark at 8″. Using a square, scribe a line across the face of the material at each mark. See Figure 72.
3. Using a straightedge and clamps as a guide for the router, remove the material between the lines across the board to a depth of ⅜″. See Figure 73.
4. Fit the top and bottom 1 x 4 x 4′-0″ back rails into the dados, flush at the ends.

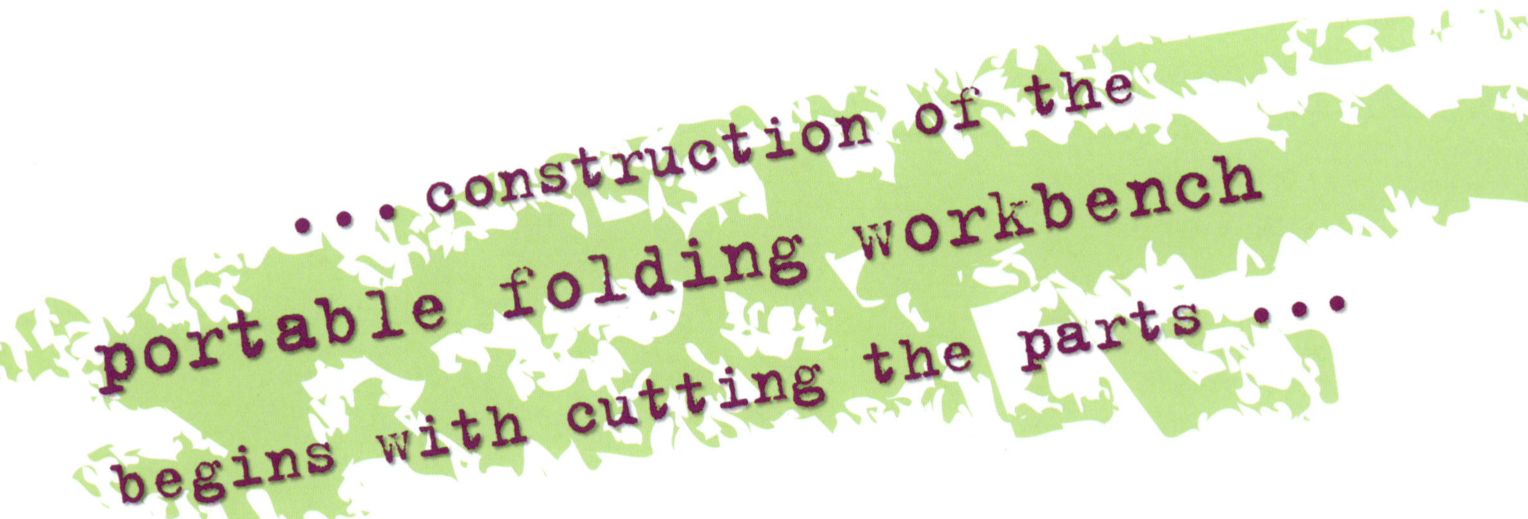

Figure 72
Figure 72 illustrates step 2

Figure 73
Figure 73 illustrates step 3

Figure 74
Figure 74 illustrates step 5

5. Fasten the back rails to the back legs with four screws at each joint. See Figure 74.
6. Fasten the 2 x 4 x 4′-0″ back face member to the back frame, flush at the ends and top of the frame, with two screws at each joint. See Figure 75.
7. Using a combination square set at ¾″, scribe a line along the face and edge of the 2 x 4 x 8′-0″ members to lay out a ¾″ x ¾″ rabbet. Cut the rabbet.
8. Glue and fasten the 1 x 4 x 3′-10 ½″ shelf members to the 2 x 4 x 4′-0″ routed top member at the routed edge. See Figure 76.

Figure 75
Figure 75 illustrates step 6

Figure 76
Figure 76 illustrates step 8

9. Glue and fasten the 1 x 4 x 3 ½″ back shelf member to the 2 x 4 x 4′-0″ routed top member at the routed edge. See Figure 77.
10. Fasten the assembled 2 x 4 x 4′-0″ routed top member to the back frame, flush with the ends and the front edge of the back frame, with screws from the top down at 12″ o.c. See Figure 78.

Figure 77
Figure 77 illustrates step 9

Figure 78
Figure 78 illustrates step 10

Figure 79
Figure 79 illustrates step 1

Assemble the Side Frames

In this procedure you will assemble the side frames using the front legs and side rail pieces. This assembly will be fastened with screws, glue, and a dowel. You will use an electric drill to make a ¾″ deep hole for the dowel.

Figure 80
Figure 80 illustrates step 3

PROCEDURE

1. Measure 4 ½″ from one end of each 1 x 4 x 2′-11″ front leg and make a mark across the face of the material using the square. See Figure 79.
2. Fasten one of the 1 x 4 x 1′-7 ½″ side rails to the front leg with the lower edge of the 1 x 4 aligned with the mark made in step 1 using four screws.
3. Fasten another side rail to the front leg, flush with the top of the leg, using four screws. See Figure 80.
4. Center mark and install strap hinges at the end of each top and bottom side rail with a screw in each hole in the leaf. See Figure 81.
5. Measure 1 ¼″ from the front edge of the 1 x 4 front leg and make a mark across the top of the leg.

Figure 81
Figure 81 illustrates step 4

Figure 82
Figure 82 illustrates step 6

6. Measure ⅜" from the edge of the leg and make a mark intersecting the line made in step 5. See Figure 82.
7. Drill a hole ¾" deep for the dowel at the mark made in step 6. See Figure 83.
8. Glue and install the dowel. See Figure 84.
9. Repeat steps 1–8 for the opposing side frame.

Figure 83
Figure 83 illustrates step 7

Figure 84
Figure 84 illustrates step 8

Join the Back Frame and Side Frames

Now the back frames and side frames can be linked with hinges. The hinges will be attached to the frames with screws.

PROCEDURE

1. Place the side frame on the inside of the 2 x 4 back leg.
2. Fasten the leaf of one hinge to the 1 x 4 lower horizontal back rail on the back frame with a screw through each hole. See Figure 85.
3. Fasten the leaf of one hinge to the 1 x 4 upper horizontal back rail on the back frame with a screw through each hole. See Figure 86.
4. Repeat steps 1–3 for the opposing side frame.

Figure 85
Figure 85 illustrates step 2

Fasten the leaf of one hinge...

Figure 86
Figure 86 illustrates step 3

Build the Tabletop

The tabletop or work surface of the bench will be built by inserting the ¾″ plywood component into the rabbet of the frame members. The plywood is then fastened to the members using construction screws.

PROCEDURE

1. Position all four tabletop frame members with the rabbetted side, previously cut, facing up and the mitered cuts aligned and tight and fasten with glue and screws. See Figure 87.
2. Insert the 14 $\frac{9}{16}$″ x 3′-6 ⅝″ x ¾″ plywood member into the rabbetted edges of the tabletop frame members.
3. Fasten the plywood to the frame members with 1″ construction screws, three in each end and four in each side, making sure the mitered joints remain tight. See Figure 88.
4. Mark and install the piano hinge along the back side of the assembled tabletop frame. See Figure 89.

Figure 87
Figure 87 illustrates step 1

Figure 88
Figure 88 illustrates step 3

Figure 89
Figure 89 illustrates step 4

Figure 90
Figure 90 illustrates step 8

See Figure 91
Figure 91 illustrates step 2

5. Attach the piano hinge to the bottom side of the 2 x 4 x 4′-0″ horizontal back face member in the back frame.
6. Lower the tabletop and adjust the side frames until they are square with the back frame.
7. Mark each location where the dowels in the side frames meet the table top.
8. Drill holes ¾″ deep at each mark to accept the dowel pins. See Figure 90.

Install the Bottom Shelf

The final stage in construction of the portable folding workbench is to install the bottom shelf. The shelf will be attached to the back rail with strap hinges.

PROCEDURE

1. Mark and install two strap hinges along the plywood bottom shelf, 8″ from each end, with ¾″ construction screws.
2. Attach the other leaf of the strap hinges to the lower horizontal back rail with ¾″ construction screws. See Figure 91.
3. Sand and finish the workbench as desired.

See Figure 92
Completed portable folding workbench

Student Name:________________________________ Date: ________________

Skateboard Ramp Project Evaluation

PROCEDURE	CRITERIA AS SPECIFIED BY THE PROJECT	POSSIBLE POINTS	SCORE
Cutting and Assembly	End cut (2)	5	
	Center support cut	5	
	Blocking cut (16)	32	
	Wedge cut (6)	12	
	Finished wedge assembly (2)	10	
	Blocking installed	5	
	Wedge assembly installed	5	
	Back support cut (3)	7.5	
	Pipe installed	5	
	Plywood back cut and installed	5	
	Ramp and top cut and installed	5	
	Ramp surface installed	5	
	Metal plate installed	5	
	Finished height	5	
	Finished width	5	
	Subtotal	*116.5*	
Overall	Joints tight	5	
	No rough edges	5	
	Screws installed per procedures	5	
	No splits and shiners	5	
	Edges flush	5	
	No excess glue	5	
	Subtotal	*30*	
General	Tool handling	5	
	Followed direction	5	
	Cleaned up	5	
	Safe work practices	5	
	Subtotal	*20*	
		Student score:	

Total Possible Points = 166.5

Suggested minimum acceptable score: 117 points or ______

Student's Signature: ______________________ Teacher's Signature: ______________________

Student Name:______________________________ Date: ______________

Adirondack Chair Table Project Evaluation

PROCEDURE	CRITERIA AS SPECIFIED BY THE PROJECT	POSSIBLE POINTS	SCORE
Cutting	Chair frame (2)	5	
	Back spacer	5	
	Leg (4)	10	
	Lower chair back	5	
	Back slat (4)	10	
	Arm support (2)	5	
	Seat slat (13)	26	
	Arm (2)	5	
	Upper chair	5	
	Front spacer	5	
	Subtotal	*81*	
Shape	Chair frame (2)	5	
	Lower chair back	5	
	Upper chair back	5	
	Back slat (4)	10	
	Arm shape (2)	5	
	Arm support (2)	5	
	Hole depth on seat slat (13)	26	
	Rounded edge of seat slats	5	
	Rounded edge of back slats	5	
	Rounded edge of arms supports	5	
	Rounded edge of arms	5	
	Subtotal	*81*	
Assembly	Finished height of arms	5	
	Finished seat width	5	
	Finished depth from face of the leg to back of the seat frame	5	
	Slat installation	5	
	Leg square to arm (2)	5	
	Arm support installation (2)	5	
	Location of front spacer	5	
	Carriage bolts installation	5	
	Screw installation	5	
	Joints tight	5	
	Edges flush	5	
	No rough edges	5	
	No splits or shiners	5	
	No excess glue	5	
	Subtotal	*70*	
General	Tool handling	5	
	Followed direction	5	
	Cleaned up	5	
	Safe work practices	5	
	Subtotal	*20*	
		Student score:	

Total Possible Points = 252

Suggested minimum acceptable score: 176 points or ______

Student's Signature: ____________________ Teacher's Signature: ____________________

Student Name:______________________________ Date: ______________

Portable Folding Workbench Project Evaluation

PROCEDURE	CRITERIA AS SPECIFIED BY THE PROJECT	POSSIBLE POINTS	SCORE
Cutting	Back legs (2)	5	
	Front legs (2)	5	
	Back rails (2)	5	
	Back face	5	
	Back top	5	
	Side rails (4)	10	
	Tabletop frame members 2 x 4 x 4′0″ (2)	5	
	Tabletop frame members 2 x 4 x 20″ (2)	5	
	Back shelf member	5	
	Left and right hand shelf members	5	
	Bottom shelf	5	
	Top insert	5	
	Subtotal	*65*	
Assembly	Back rails attached to back legs	5	
	Back face member	5	
	Top shelf assembled and attached	5	
	Side frame	5	
	Side rail squared to front leg	5	
	Side frame fastened to back leg	5	
	Plywood installed in tabletop frame	5	
	Tabletop assembly fastened together	5	
	Tabletop assembly attached to back assembly	5	
	Bottom shelf attached to back frame	5	
	Subtotal	*50*	
Overall	Height	5	
	Width	5	
	Depth	5	
	Edges flush	5	
	Joints tight	5	
	Screws installed per procedures	5	
	Dowel holes in tabletop	5	
	No splits and shiners	5	
	No rough edges	5	
	No excess glue	5	
	Subtotal	*50*	
General	Tool handling	5	
	Followed direction	5	
	Cleaned up	5	
	Safe work practices	5	
	Subtotal	*20*	
		Student score:	

Total Possible Points = 185

Suggested minimum acceptable score: 130 points or ______

Student's Signature: ____________________ Teacher's Signature: ____________________

CHAPTER 6

SHED

CONTENTS

Introduction

Your earlier carpentry projects involved mostly construction of picnic tables, outdoor chairs, sawhorses, or workbenches. While working on these worthwhile projects you learned a number of common carpentry techniques. You also developed basic skills in the handling and application of common carpentry tools, materials, and fasteners. In this chapter you will turn your attention to building stand-alone structures. This will enable you to learn more advanced carpentry techniques and skills.

Figure 1
Frame

Stand-alone structures, such as sheds, are built using a frame, as shown in Figure 1. These framing members are like the bones of the house; they give it its shape and structure. Sheets of material called sheathing are nailed over the frame, as shown in Figure 2. The sheathing can be compared to the skin of the house in that it encloses it and keeps it all together. Sheathing also provides a great deal of the strength to the structure.

Figure 2
Sheathing over framing

Figure 3
Floor framing

Structures are made up of many different framing members, and every section of a structure has its own type of framing. Floors, walls, and ceilings all use different types of framing members. The floor framing, shown in Figure 3, supports the wall framing members, which are aligned precisely over the floor framing members. The wall framing supports the roof framing members, which are aligned precisely over the wall framing members. Each section of framing relies on the one below it for support.

Because of this, the location of framing members is critical. Detailed drawings are created showing the exact location of each framing member. Before construction, the location of each framing member is laid out on the actual lumber used to frame the structure, as shown in Figure 4.

This chapter includes detailed instructions for laying out, framing, and building a shed with a gable roof, a garden tool shed, and a wishing well. Each of these projects will provide you with additional experience in standard carpentry processes such as print reading, measurement, layout, cutting, and assembling. You will also get additional practice with the safe handling and use of hand and power tools.

Figure 4
Lumber laid out

What's New?

Most of the tools and materials you will be using were introduced to you in earlier projects. These include hand tools such as hammers, hand saws, chisels, and framing squares and power tools such as circular saws, electric drills, and screw guns. They also include materials and fasteners such as lumber, plywood, and construction screws. This chapter will introduce you to additional carpentry tools and types of carpentry material and show you new ways of using material that you have used before. The projects in this chapter come a step closer to some of the actual tasks performed by a professional carpenter. Some of the tools and materials that may, as yet, be unfamiliar to you include the following:

Aviation Snips Aviation snips are a cutting tool that is used for cutting light gauge metal. See Figure 5.

Figure 5
Aviation snips

Cane Bolt An L-shaped piece of door hardware, a cane bolt is a vertical locking device that secures a door by sliding into a recess in the floor. See Figure 6.

Figure 6
Cane bolt

The **projects** in this chapter come a step closer to some of the **actual tasks** performed by a professional **carpenter.**

Figure 7
Dovetail saw

Figure 8
Galvanized flat strapping

Dovetail Saw Used to make narrow cuts, a dovetail saw has a straight handle, a reinforced back, and fine teeth. See Figure 7.

Galvanized Flat Strapping Galvanized flat strapping is a flexible metal strap, ranging from 1/2″ to 1 1/2″, typically perforated to accept fasteners. See Figure 8. It is used for several purposes including reinforcing framing and is treated with zinc to prevent rust.

Galvanized Nails Sometimes nails are coated with zinc to prevent rust. See Figure 9. These are called galvanized nails, and they are often used as fasteners on projects likely to be exposed to moisture.

Figure 9
Galvanized nails

Figure 10
Hasp

Hasp A hasp is a hinged metal device with a slot designed to accept a bolt or lock. See Figure 10.

On Center Layout Carpenters space framing parts at regular intervals on layouts. The spacing often used is 16″ on center. This means that from one end of the board to the center of the framing parts is a space of 16″ multiples, such as 16″, 32″, 48″, and so on. Most retractable tape measures indicate the 16″ on center marks with an arrow, as shown in Figure 11, or some other specific symbol.

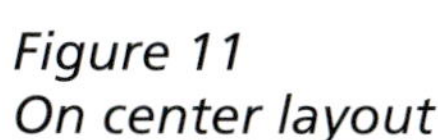

Figure 11
On center layout

Joist clip angle

Figure 12
Joist clip angles

Joist Clip Angles Joist clip angles are L-shaped metal fasteners that are used in floor and wall framing. See Figure 12.

Paling A row of upright pointed sticks forming a fence is called paling; each stick is referred to as a pale. See Figure 13.

Sheathing Sheets of plywood, OSB, or other material nailed to framing are called sheathing.

Figure 13
Paling

Sinker When driving fasteners below the surface of wood, carpenters sometimes use a type of nail called a sinker. Thinner than a common nail, a sinker has a funnel-shaped head with a grid pattern stamped on it. Sinkers are coated with an adhesive to provide greater holding power.

Tie Metal ties are used to brace carpentry joints. A type of tie called a hurricane tie is shown in Figure 14. Usually, they consist of a flat or L-shaped plate that is hammered or screwed in place.

Metal ties are used to brace carpentry joints.

Figure 14
Hurricane tie

1 Layout

Before building any of the framing for a project, framing members must be cut to the dimensions in the drawings and then marked at the locations where they will be fastened together. The process of marking the framing members is called layout. Accurate layout is critical to the success of any structure.

Before assembly, framing members should be visually inspected for structural defects. Most lumber has a slight natural arc along the edge created by the grain of the wood, as shown in Figure 15. This curve is called a crown. Lumber that is used for framing members should be assembled with the crown up. This is because the crown will tend to straighten as weight is placed on the framing. This section includes practice exercises on how to mark the location of framing members on a 16″ on center (o.c.) layout and how to adjust layout for the thickness of a wall.

Figure 15
Crown

Marking a 16″ On-Center Layout

You will need the following materials:

- Scrap 1 x or 2 x material 8′-0″ or longer
- 6d nail

You will need the following tools:

- Tape measure
- Speed square
- Pencil
- Hammer
- (2) Sawhorses

Figure 16
Figure 16 illustrates step 1

LEAD-UP EXERCISE

1. Measure and mark 15 ¼″ from the end of the material and partially drive a nail at that mark. See Figure 16.
2. Hook the tape measure on the partially driven nail and mark the material at 16″ and then make an "X" after the mark. See Figure 17.
3. Continue marking the material at every 16″ and making Xs, leaving the tape measure hooked to the nail. You will be making a mark at 32″, 48″, 64″, and so forth, with an X after each mark.
4. Square lines at the marks with the speed square across the material making sure that an X is placed after each mark.

Figure 17
Figure 17 illustrates step 2

Adjusting the Layout for Wall Thickness

In some cases you will not be starting your layout from the end of the board. You may have to adjust for wall thickness. Instead of hooking the tape measure on the end of the material, it must be extended beyond the end by the thickness of the wall. In the following procedure the wall thickness is assumed to be 3 ½″.

You will need the following materials:

- Scrap 1x or 2x material 8′-0″ or longer
- 6d nail

You will need the following tools:

- Retractable tape measure
- Speed square
- Pencil
- Hammer
- (2) Sawhorses

Figure 18
Figure 18 illustrates step 1

LEAD-UP EXERCISE

1. Place your tape measure with the 3 ½″ mark at the end of the material. See Figure 18.
2. Measure and mark 15 ¼″. See Figure 19.
3. Partially drive a nail at the mark made in step 2. See Figure 20.
4. Hook the tape measure on the partially driven nail and mark the material at 16″ and then make an "X" after the mark.
5. Continue marking the material at every 16″ and making Xs, leaving the tape measure hooked to the nail. You will be marking 32″, 48″, 64″ and so forth, with an X after each mark.
6. Square lines at the marks with the speed square across the material making sure that an X is placed after each mark.

Figure 19
Figure 19 illustrates step 2

Figure 20
Figure 20 illustrates step 3

2 Floor Framing

The floor framing supports the weight of the entire structure. Because of this, it is especially important that it is laid out and constructed accurately and that the floor is square when it is finished. This does not mean that the floor is shaped like a square. It means that all four corners are 90°.

There are three basic parts to floor framing. They are floor joists, rim joists, and sheathing, as shown in Figure 21.

Floor Joists Horizontal framing members of a floor that directly support the sheathing and the weight of the structure are called floor joists.

Rim Joists Rim joists are nailed along the ends of the floor joists to prevent them from twisting or tipping. Both the floor joists and the rim joists are made from the same size lumber.

Figure 21
Floor framing components

Sheathing Panel material, typically plywood or OSB, that is installed on top of the joists is called sheathing. It is also used in wall and roof framing. Sheathing is an important part of the frame of a structure. It helps tie the floor joists and rim joists together and increases the rigidity of the floor. When installing sheathing, the sheets are placed so that edges meet in the center of the framing members.

It is important that the floors are framed square because the floor is the basis for the rest of the building. If the floor is out of square, the errors may be magnified in the wall framing and in the roof framing. One method of checking for square is using the diagonal measurement method. If a framing assembly, such as a floor or wall, is square, the diagonal measurements from corner to corner will be the same providing that the opposite sides are equal in length.

Checking for Square Using the Diagonal Method

You will need the following material:

- 4′-0″ x 8′-0″ plywood

You will need the following tools:

- Tape measure
- (2) Sawhorses

Figure 22
Figure 22 illustrates step 1

LEAD-UP EXERCISE

1. Measure from one corner diagonally to the other corner. See Figure 22.
2. Write down the measurement.
3. Measure the two opposite corners diagonally. See Figure 23.
4. Compare the two measurements. The measurements should be the same.

Figure 23
Figure 23 illustrates step 3

3 Wall Framing

There are several types of wall framing members. They include vertical members called studs and horizontal members called plates. There are also framing members for windows and doors called headers and rough sills. See Figure 24.

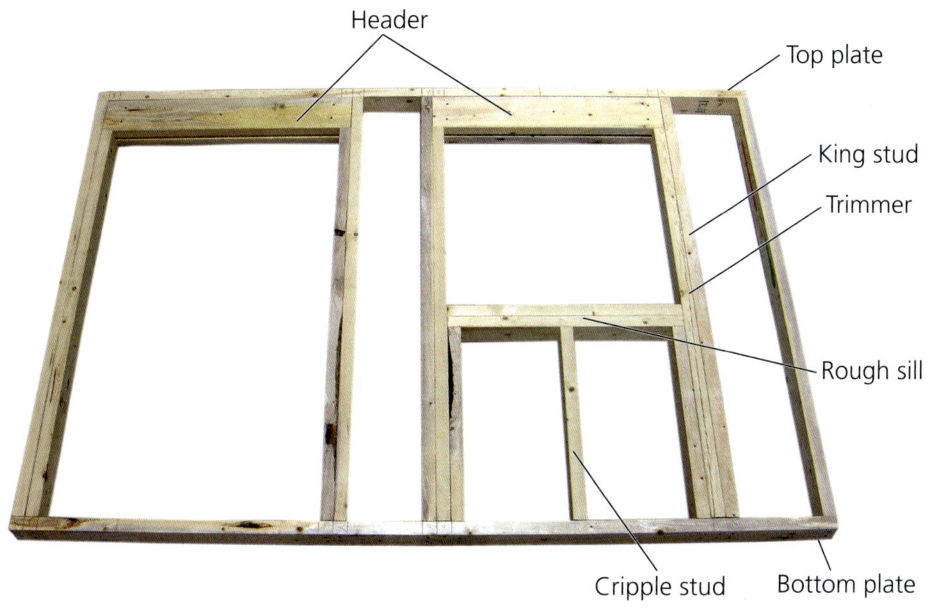

Figure 24
Wall framing components

Vertical framing members, usually called studs, include the following different types:

Common Studs Common studs run the full height of the wall between the top and bottom plates.

King Studs A king stud is a common stud that is used at window and door openings.

Trimmers Shortened studs used to support headers are called trimmers. They are fastened to full-length studs, called king studs, that are fastened to the header on either side of the opening.

Cripple Studs Shorter studs that have been cut to allow for an opening are called cripple studs. They extend from the plate to the underside of the rough sill and from the header to the top plate.

Stud-Block-Stud Corner A stud-block-stud corner is a corner built from two common wall studs with blocking fastened between them. The blocking consists of short pieces of wood that are used to space the framing members.

In addition to studs, wall framing includes the following horizontal framing members:

Plates The plates are horizontal framing components that form the tops and bottoms of walls. The studs are attached to the top and bottom plates and together the studs and plates form the wall frame. When necessary to support ceiling joists, rafters, or other members, the top plate is doubled to strengthen, align, and interlock walls. This is called a double top plate.

...most walls are assembled on the ground and then raised into positi

Headers Horizontal framing components that form the top of rough wall openings are called headers. A rough opening is the framed opening in a wall into which a finished door or window unit will be placed. A built-up header is made from plywood spacers and 2x material.

Sills Sills form the bottom of a rough window opening and are used to secure the top ends of the cripple studs.

Most walls are assembled on the ground and then raised into position, as shown in Figure 25. The walls in all of these projects are heavy and will require at least two people to raise them. Trying to do it by yourself is not safe. Find someone to help you. The procedures in this section will detail the steps for assembling, squaring, standing, and bracing a wall and how to do it safely.

Figure 25
Raising a wall

Constructing a Built-Up Header Using Plywood for Standard 2 x 4 Construction

You will need the following material:

- (2) 2 x 4 x 3′-0″ lumber
- (4) ½″ x 2″ x 3″ plywood or OSB for spacers
- 6d nails
- 16d nails

You will need the following tools:

- Hammer
- Circular saw
- Tape measure
- Speed square
- Pencil
- Safety glasses, hard hat, work shoes, and other PPE
- (2) Sawhorses

LEAD-UP EXERCISE

1. Crown the 2 x 4s and identify the crowns with arrows.
2. Orient the 2 x 4s so the crowns are facing in the same direction.
3. Create spacers by ripping 2″ strips of ½″ sheathing material. Cut the strips to a length of ½″ less than the width of the header material. See Figure 26.
4. Nail the spacers to the inside face of the header material with 6d nails. Locate the first two pieces 2″ from each end. Place remaining pieces 16″ o.c.

Figure 26
Header spacers

5. Pair up the two pieces of 2x material, matching the crowns and aligning the ends and edges.
6. Nail through the 2x material and spacers at the spacer locations with 16d nails. It is a good practice to nail through both faces of the header. See Figure 27 for an example of a completed built-up header.

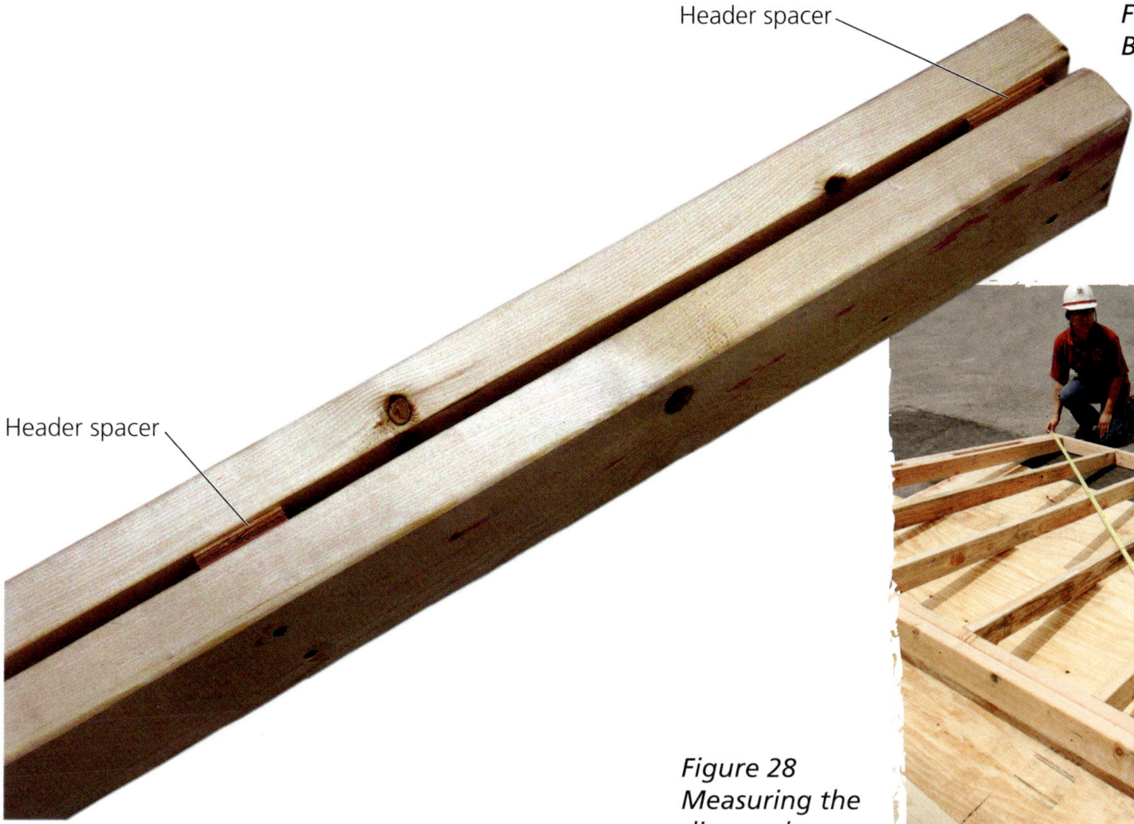

Header spacer

Header spacer

Figure 27
Built up header

Figure 28
Measuring the diagonal

Squaring Walls

Hold the bottom plate straight and in a fixed position. Check for square by measuring the diagonals of the wall. Measure from the outside corner of the bottom plate to the outside corner of the top plate, as shown in Figure 28. If diagonals are not equal dimensions, it will be necessary to move the wall toward the shorter dimension. Move the wall approximately half the difference of the two diagonal dimensions. This is called racking the wall. Install a temporary brace approximately 45° diagonally across the face of the wall, as shown in Figure 29. To hold the wall square, attach the brace to the top plate, bottom plate, and intermediate studs.

Figure 29
Installing a temporary brace

4 Roof Framing

There are many different kinds of roofs, and each one is framed differently. Most of them, however, use some of the same framing members. The equal slope gable roof is one of the more basic types of roof to build because it slopes equally in only two directions and has only one type of rafter. The main components of the gable roof are shown in Figure 30 and include the ridge board and common rafters.

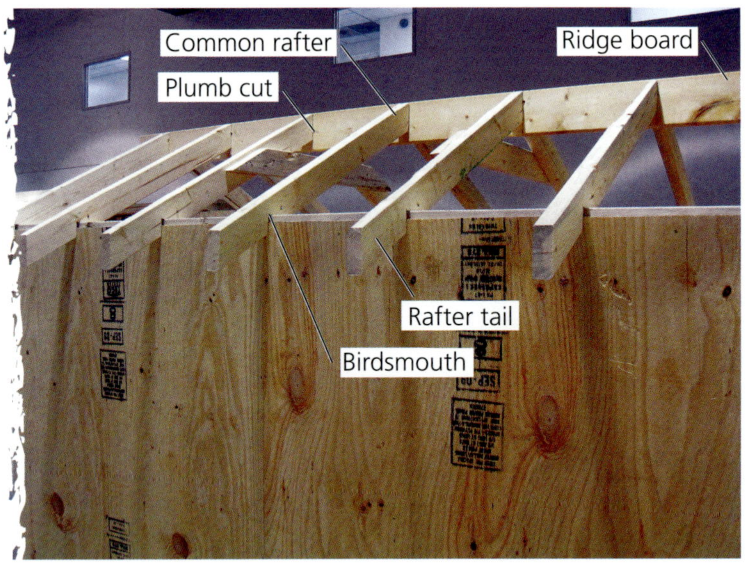

Figure 30
Roof framing components

Common Rafter The common rafter is a roof member that runs from the outside of the wall to the ridge board, perpendicular to, and bearing on, the double top plate. A fly rafter is a common rafter that is installed outside the building line, which is the outside edge of the exterior wall sheathing.

Ridge Board The ridge board is a horizontal framing member that is located at the top of the roof and is the board to which the rafters are fastened. The ridge is the highest point of the roof.

Standard terms are used to identify various parts of the rafter during layout. Starting at the upper end of the rafter is a plumb cut at the ridge board.

Plumb Cut Any vertical cut made on a rafter is a plumb cut. Plumb cuts are made at the ridge, birdsmouth, and often at the rafter tail.

Birdsmouth The birdsmouth is a notch cut in the rafter that is used to lock the rafter over the double top plate on the outside wall. The birdsmouth includes a plumb cut and level cut.

Seat Cut The level cut of the birdsmouth is called the seat cut. A level cut is any horizontal cut on the rafter usually at the birdsmouth and rafter tail.

Rafter Tail The part of the rafter that extends beyond the outside edge of the exterior wall is the rafter tail.

Positioning Rafter Material for Layout During rafter layout, the rafter material is placed across two sawhorses with the crown of the material toward you. When placed in this position, rafter layout is performed from left to right, which is from the ridge to the rafter tail.

Orienting the Framing Square To begin rafter layout, the framing square is held with the tongue of the square in the left hand and the body or blade of the square in the right hand. With this orientation, the corner of the framing square will be pointing away from you.

Laying Out and Cutting Rafters A tape measure is hooked on the ridge end of the rafter material and the line length and total line length are measured and marked. After marking plumb lines on the rafter at the ridge, building line, and tail, the seat cut is laid out. Finally, a plumb cut at the tail is cut.

Cutting a Common Rafter Pattern

You will need the following material

- (1) 2 x 4 x 8′-0″ lumber

You will need the following tools

- Tape measure
- Framing square
- Pencil
- Circular saw
- Hand saw
- (2) Sawhorses

LEAD-UP EXERCISE

1. Place the 2 x 4 x 8′-0″ rafter material on the sawhorses with the crown toward you.
2. Hook the tape measure on the left end of the rafter material, along the crowned edge and mark 2′-6 7⁄8″ and 3′-7 1⁄2″. See Figure 31.

Figure 31
Figure 31 illustrates step 2

3. Position the framing square at the left end on the crown side of the rafter material, as shown in Figure 32, holding 4″ on the tongue of the square and 12″ on the body of the square.
4. Scribe along the tongue side of the square, as shown in Figure 33.

Figure 32
Figure 32 illustrates step 3

Figure 33
Figure 33 illustrates step 4

Figure 34
Figure 34 illustrates step 5

5. Repeat steps 3 and 4 at the 2′-6 ⅞″ mark and the 3′-7 ½″ mark, as shown in Figure 34.
6. Measure 2 ½″ along the 2′-6 ⅞″ plumb line and mark this point. See Figure 35.
7. Scribe a square line from the 2 ½″ mark made in step 6 for the seat cut. See Figure 36.

Figure 35
Figure 35 illustrates step 6

Figure 36
Figure 36 illustrates step 7

Figure 37
Figure 37 illustrates a completed step 8

8. Cut to the lines with the circular saw. Figure 37 shows a completed cut.
9. Finish the birdsmouth cut with a hand saw. See Figure 38. Figure 39 shows a completed birdsmouth.
10. Cut the line scribed in step 4 and the line scribed at the 3′-7 ½″ mark made in step 5.

Figure 38
Figure 38 illustrates step 9

Figure 39
Completed birdsmouth

5 Shed with a Gable Roof

The focus of this project is on framing the shed. The framing is basically the same as a professional carpenter would use to build a house, but on a smaller scale. In this project, you will frame a floor, four walls, and a gable roof. A gable roof is a roof that slopes equally in only two directions and has only one type of rafter. You will also make a window opening in one wall and a door opening in another wall. When carpenters build a house, they usually frame the openings for the windows and doors and then install the window and door units into the openings later.

Note: This project has procedures that may require more than one person.

You will need the following materials

Floor System:

- (2) ¾″ x 4 x 8 plywood
- (2) 2 x 6 x 8′-0″ rim joists
- (4) 2 x 6 x 10′-0″ floor joists

Wall system:

- (6) 2 x 4 x 8′-0″ top/bottom plates
- (3) 2 x 4 x 10′-0″ top/bottom plates
- (36) 2 x 4 x 8′-0″ studs/headers
- Scrap ½″ plywood for headers

Roof system:

- (1) 2 x 6 x 8′-0″ ridge board
- (7) 2 x 4 x 8′-0″ rafters

Temporary braces:

- (6) 2 x 4 x 10′-0″

Fasteners:

- 12d sinkers
- 6d common nails

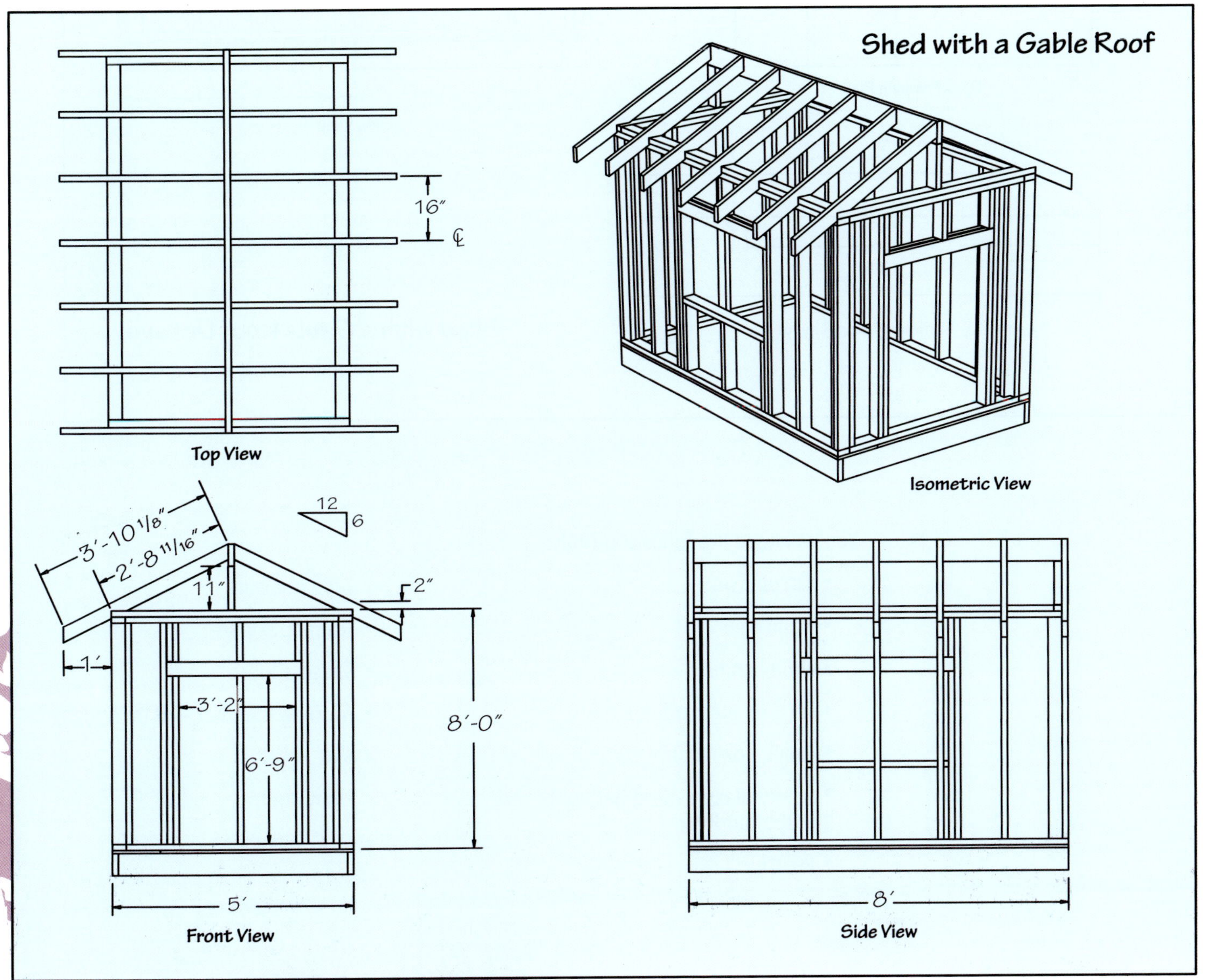

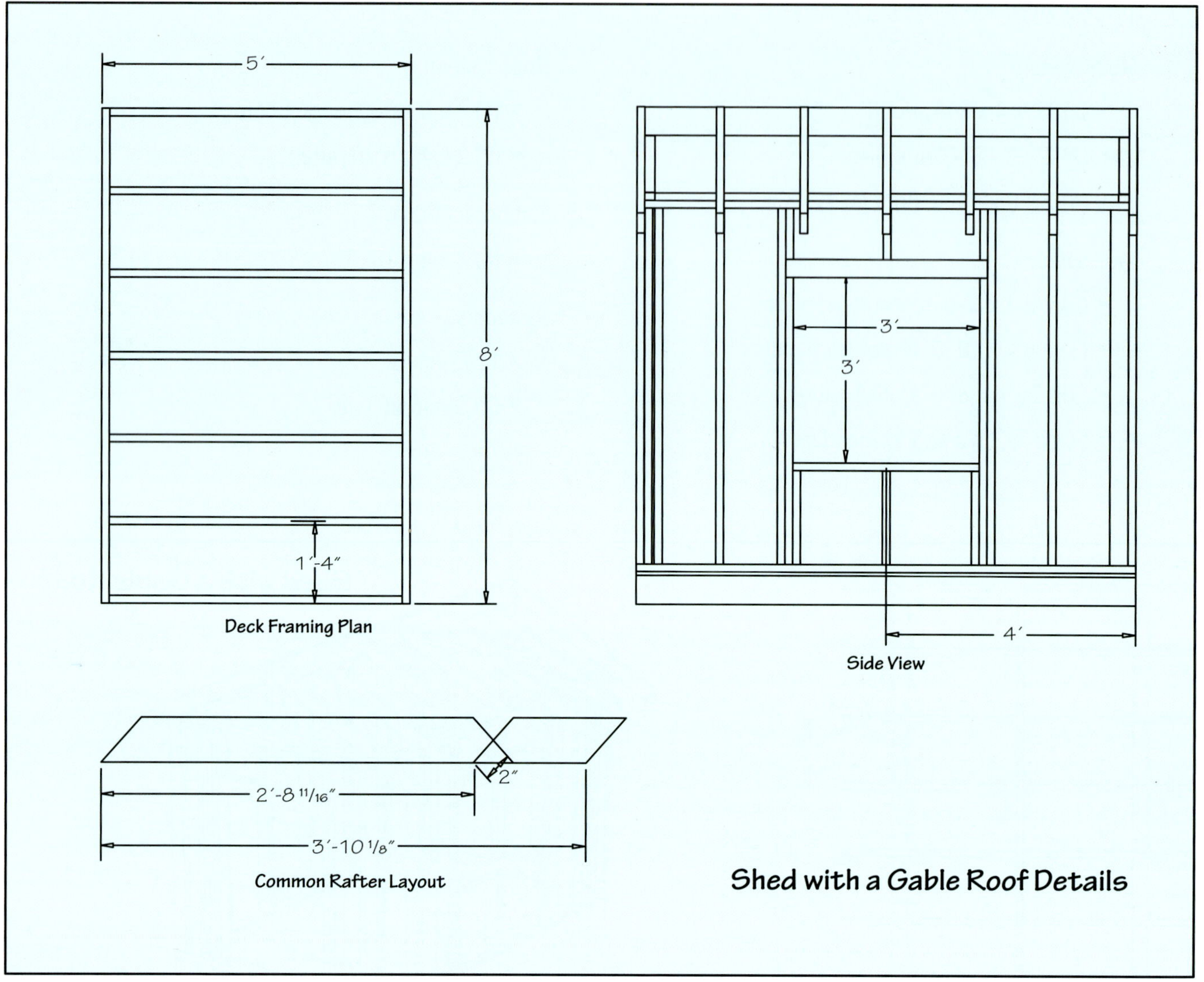

You will need the following tools:

- Circular saw
- Hand saw
- Utility knife
- (2) 8′-0″ step ladders
- Chalk box/chalk
- Safety glasses, hard hat, work shoes, and other PPE
- Pencil
- Tape measure
- Framing square
- Speed square
- Framing hammer—16 to 22 oz
- 4′-0″ level
- Plumb bob
- (2) Sawhorses

Frame the Floor System

The first step in building the shed is to frame the floor. It is very important to build the floor carefully so it is strong and square. If the floor is square, it will be much easier to make the rest of the shed square.

PROCEDURE

1. Check both 2 x 6 x 8′-0″ rim joists for length and square. Cut if necessary.
2. Turn both rim joists on edge with the crowns up and temporarily fasten them together with 8d sinkers. See Figure 40.
3. Measure and mark 15 ¼″ from the end of the rim joist, shown in Figure 41, and drive a nail at that mark.

Figure 40
Figure 40 illustrates step 2

Figure 41
Figure 41 illustrates step 3

Figure 42
Figure 42 illustrates step 4

4. Hook the tape measure on the nail and mark the rim joist every 16". Make an X after each mark. You will make a mark at 32", 48", 64", and so forth, with an X after each mark. See Figure 42.
5. Transfer the layout marks with the speed square across both rim joists making sure that an X is placed on both rim joists. See Figure 43.
6. Separate the rim joists.
7. Square one end of a 2 x 6 x 10′-0″ board. Cut if necessary.
8. Measure 4′-9″ from the square end and cut a floor joist and repeat for the second joist.
9. Repeat steps 8 and 9 for the remaining 5 floor joists.
10. Align the floor joists on the 16″ o.c. layout marks between each rim joist with the crown up.

Figure 43
Figure 43 illustrates step 5

11. Face nail the rim joist to the floor joist ends using 12d sinkers. See Figure 44.
12. Check the floor system for square using the diagonal method. See Figure 45. Adjust as needed.
13. Place the sheet of ¾″ plywood on top of the floor framing perpendicular to the floor joist.

Figure 44
Figure 44 illustrates step 11

Figure 45
Figure 45 illustrates step 12

Figure 46
Figure 46 illustrates step 15

14. Align the edges of the sheet with the rim joist and floor joists.
15. Nail both corners of the plywood to the rim joist with a 6d common nail in each corner. See Figure 46.
16. Recheck the floor system for square using the diagonal method. See Figure 47.
17. Nail the plywood to the floor joist at the opposite corner with a 6d common nail. Make sure that the outside face of the floor joist is flush with the edge of the plywood sheet.

Figure 47
Figure 47 illustrates step 16

18. Nail across the length of along the rim joist and at each floor joist location at the opposite edge of the plywood. See Figure 48.
19. Measure and then rip a piece of plywood to finish sheathing the floor system See Figure 49.
20. Nail the plywood to the floor framing with 6d common nails.

Figure 48
Figure 48 illustrates step 18

Figure 49
Figure 49 illustrates step 19

Frame the Side Wall Without the Window

In building this first wall, you will cut a top plate, bottom plate, nine common wall studs and two stud-block-stud corners. You will build the wall on the floor and stand it into place once the wall is finished. For this process, you will need someone to help you.

PROCEDURE

1. Check two 2 x 4 x 8′-0″ boards for length and cut them if necessary. These are the top and bottom plates for one side wall.
2. Tack the top and bottom plate together at each end with 6d common nails.
3. Lay out and mark the wall stud locations matching the floor joists layout.
4. Cut nine 2 x 4 x 8′-0″ boards 7′-7 ½″ long for common wall studs and corner studs.
5. Build two stud-block-stud corners. See Figure 50.

Figure 50
Figure 50 illustrates step 5

Figure 51
Figure 51 illustrates step 8

6. Separate the top and bottom plates.
7. Place the corners flush with the plate ends.
8. Attach the top and bottom plates to the stud-block-stud corner assemblies with 12d sinkers. See Figure 51.
9. Place the wall studs on layout between the top and bottom plates.
10. Nail the wall studs to the top and bottom plates using two 12d sinkers per stud per plate. See Figure 52.

Figure 52
Figure 52 illustrates step 10

11. Cut one 2 x 4 x 8′-0″ board to 7′-5″ long for the double top plate.
12. Measure and make a mark 3 ½″ from one end of the top plate and align the double top plate to this line. Install the double top plate with 12d sinkers.
13. Check the wall frame for square by measuring the diagonals. Adjust the frame as needed. See Figure 53.
14. Install a temporary brace to keep the wall square. See Figure 54.

Figure 53
Figure 53 illustrates step 13

Figure 54
Figure 54 illustrates step 14

Figure 55
Figure 55 illustrates a completed step 15

15. Stand the wall, aligning the edge of the bottom plate with the outside edge of the rim joist and flush with both ends of the floor. See Figure 55. This will take at least two people.
16. Secure the bottom plate at each end and at the center with 12d sinkers.
17. Install a temporary brace to keep the wall upright. See Figure 56.

Install a temporary brace...

Figure 56
Figure 56 illustrates step 17

Figure 57
Figure 57 illustrates step 4

Frame the Side Wall with the Window

This wall is more complex than the first wall because you must frame it for a window. In addition to cutting common wall studs and stud-block-stud corners, you will make and install king studs, trimmer studs, cripple studs, rough sill, and a header. When the wall is finished, you will need someone to help you stand it into place.

PROCEDURE

1. Check two 2 x 4 x 8′-0″ boards for length and cut if necessary. These are the top and bottom plates for one side wall.
2. Tack the top and bottom plate together at each end with 6d common nails.
3. Locate the window opening on the drawing.
4. Locate and mark the center line for the window at 4′-0″ on the top and bottom plate, and mark with the symbol for centerline, which is the letter C superimposed on the letter L. See Figure 57.
5. Divide the rough opening width of 3′-0″ in half.
6. Mark 18″ on one side of the centerline. See Figure 58. Mark 18″ on the other side of the centerline.

Figure 58
Figure 58 illustrates step 6

Figure 59
Figure 59 illustrates step 7

7. Square a line across both plates at both marks. See Figure 59. These lines represent the inside edge of the trimmer studs. Place the letter T on the outside of the lines to locate the trimmer studs.
8. Mark the letter C on the inside of both lines made in step 7 to locate the cripple studs. See Figure 60.
9. Measure 1 ½″ from the lines made in step 6 for the trimmer studs. See Figure 61. Square lines across both plates and mark an X on the outside of the lines. This represents the king stud location.
10. Lay out the wall stud locations matching the floor joist layout.
11. Cut eight 2 x 4 x 8′-0″ boards 7′-7 ½″ long for common wall studs, corner studs, and king studs.
12. Build two stud-block-stud corners.
13. Cut and build a 3′-3″ long header using a 2 x 4 x 8′-0″ and scrap ½″ plywood spacers.

Figure 60
Figure 60 illustrates step 8

Figure 61
Figure 61 illustrates step 9

14. Separate the top and bottom plates.
15. Align the stud-block-stud corners flush with the ends of the top and bottom plates.
16. Attach the top and bottom plates to the stud-block-stud corner assemblies with 12d sinkers.
17. Place the wall and king studs on layout between the top and bottom plates.
18. Nail the wall and king studs to the top and bottom plates using two 12d sinkers per stud per plate.
19. Cut two 2 x 4 x 8′-0″ boards 6′-9″ long for the trimmer studs.
20. Nail the trimmer studs to the inside of the king studs and through the bottom plate with 12d sinkers.

Figure 62
Figure 62 illustrates step 21

21. Place the header built in step 13 above the trimmer studs and attach it to the king studs on each side with 12d sinkers. See Figure 62.
22. Measure, cut, and install the cripple studs above the header as marked on the layout. See Figure 63.

Figure 63
Figure 63 illustrates step 22

Figure 64
Figure 64 illustrates step 23

23. Measure, cut, and install 2 x 4 x 3′-7 ½″ cripple studs for the rough sill and attach to the bottom plate and trimmer studs as marked on the layout. See Figure 64.
24. Cut and install the 2 x 4 x 3′-0″ rough sill on top of the cripple studs matching the layout on the bottom plate. See Figure 65.

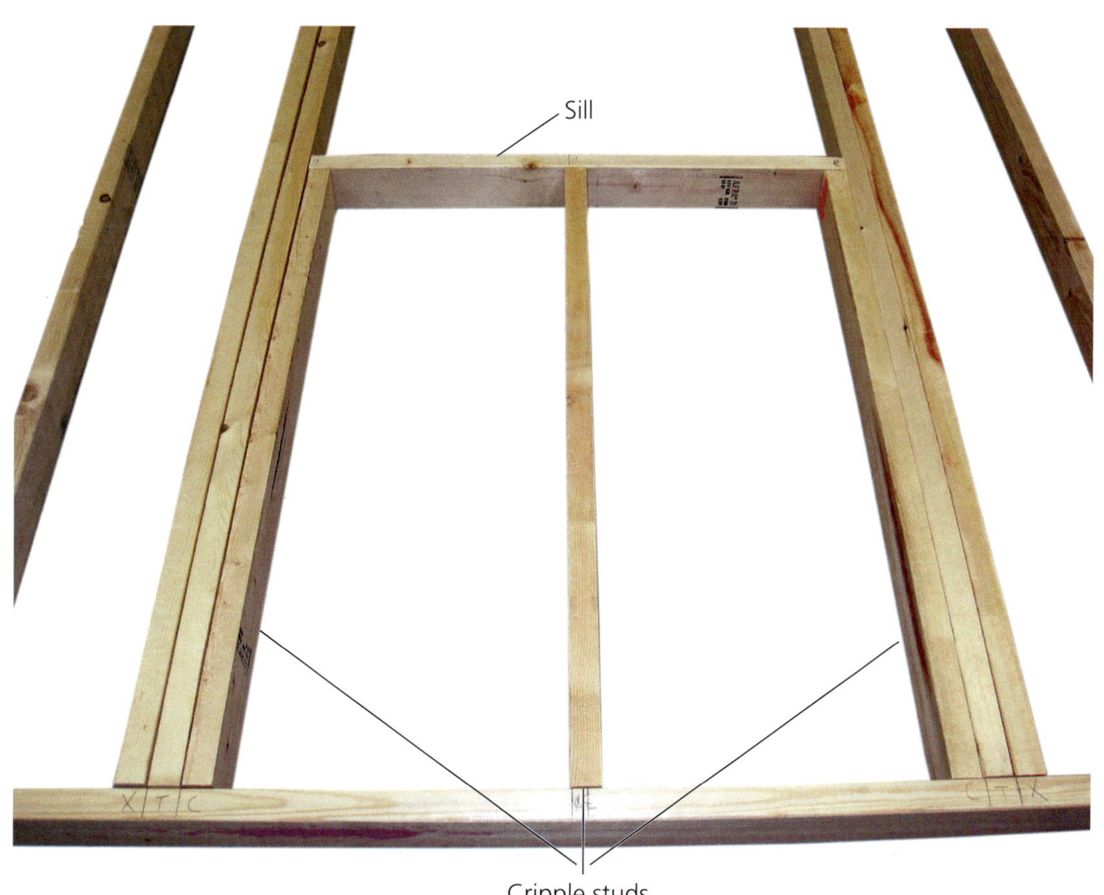

Figure 65
Figure 65 illustrates step 24

Figure 66
Figure 66 illustrates step 28

25. Cut one 2 x 4 x 8′-0″ board 7′-5″ long for the double top plate.
26. Measure and make a mark 3 ½″ from one end of the top plate and align the double top plate to this line. Attach the double top plate with 12d sinkers.
27. Check the wall frame for square by measuring the diagonals. Adjust as needed.
28. Install a temporary brace to keep the wall square. See Figure 66.
29. Stand the wall, aligning the edge of the bottom plate with the outside edge of the rim joist and flush with both ends of the floor. This will take at least two people.
30. Secure the bottom plate at each end and the center with 12d sinkers.
31. Install a temporary brace to keep the wall upright. See Figure 67.

Figure 67
Figure 67 illustrates step 31

Frame the Front Wall with the Door

Because this wall will have an opening for a door, it will be constructed in a similar way to the window wall, only without a window sill. When the wall is finished, you will need someone to help you stand it in place. Once the wall is up, you must remember not to nail the sill plate to the floor between the king studs for the door. This section of sill plate will be removed later.

PROCEDURE

1. Cut one 2 x 4 x 10′-0″ board into two 4′-5″ lengths for the top and bottom plates.
2. Tack the top and bottom plate together at each end with 6d common nails.
3. Locate the door opening on the drawing.
4. Locate the center line for the door opening on the top and bottom plate at 2′-2 ½″, and mark it with the centerline symbol. See Figure 68.

Figure 68
Figure 68 illustrates step 4

5. Divide the rough opening width of 3′-2″ in half. Mark 19″ on one side of the centerline. See Figure 69. Mark 19″ on the other side of the centerline.

Figure 69
Figure 69 illustrates step 5

Figure 70
Figure 70 illustrates step 6

6. Square a line across both the top and bottom plates at both marks. These lines represent the inside edge of the trimmers. Mark the letter T on the outside of the lines. See Figure 70.
7. Mark the letter X across the top plate on the inside of both lines made in step 6. This will be the location of the blocks over the header.
8. Measure 1 ½″ out from the lines made in step 6 for the trimmer studs and square lines across both plates. Mark an X on the outside of the lines. This is the location of the king studs.

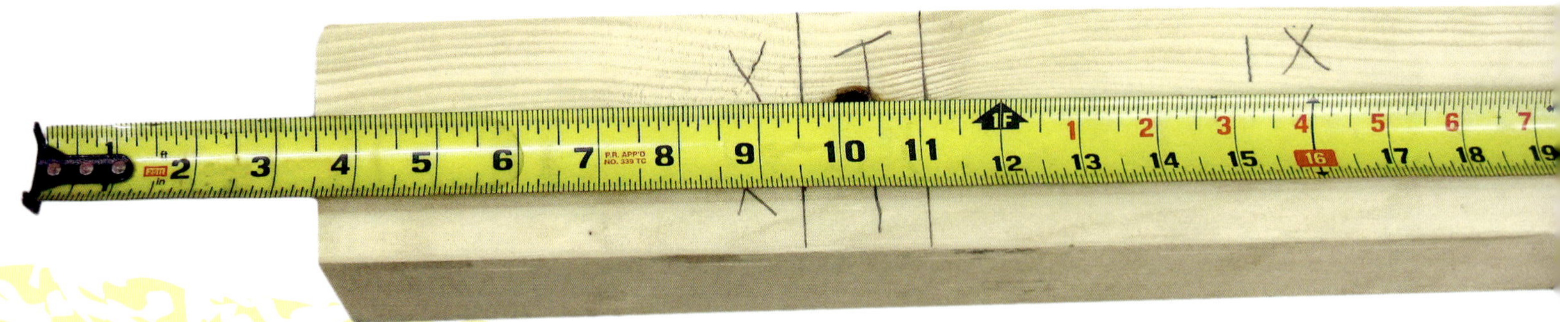

Figure 71
Figure 71 illustrates step 9

Figure 72
Figure 72 illustrates step 16

9. Lay out and mark the wall stud locations across both the top and bottom plates allowing 3 ½″ for the side wall. See Figure 71.
10. Cut four 2 x 4 x 8′-0″ boards 7′-7 ½″ long for common wall studs and king studs.
11. Cut and build a 41″ long header using a 2 x 4 x 8′-0″ board and scrap ½″ plywood spacers.
12. Separate the top and bottom plates.
13. Place wall and king studs on layout.
14. Nail the wall and king studs to the top and bottom plates using two 12d sinkers per stud per plate.
15. Cut two 2 x 4 x 8′-0″ boards 6′-9″ long for the trimmer studs.
16. Install the trimmer studs to the inside of the king studs and the bottom plate with 12d sinkers. See Figure 72.
17. Install the header made in step 11 above the trimmer studs. Attach the header to the king studs on each side with 12d sinkers.
18. Measure, cut, and install the cripple studs above the header as marked on the layout. See Figure 73.

Figure 73
Figure 73 illustrates step 18

Figure 74
Figure 74 illustrates step 21

Figure 75
Figure 75 illustrates step 23

19. Cut one 2 x 4 x 8′-0″ board 5′-0″ long for the double top plate. This piece will be installed after the wall is raised.
20. Check the wall frame for square by measuring the diagonals and adjust as needed.
21. Install a temporary brace to keep the wall square. See Figure 74.
22. Stand the wall, aligning the edge of the bottom plate with the outside edge of the floor joist. This will take at least two people.
23. Secure the bottom plate at each end and next to the king studs with 12d sinkers through the plywood and into the end joist. See Figure 75. Do not put nails in the bottom plate of the door opening.
24. Align the front wall with the two side walls and fasten the double top plate cut in step 19 to the top plates of the two side walls with 12d sinkers. See Figure 76.
25. Nail the end studs of the front wall to the stud block stud corners of the side walls. Place the nails approximately 16″ apart.
26. Remove the temporary braces holding the side walls up but leave the temporary braces holding the walls square.

Figure 76
Figure 76 illustrates step 24

Frame the Back Wall

Now that you have experience with more complex walls, this last wall will be relatively easy. As with the other walls, you will need someone to help you stand it in place.

PROCEDURE

1. Cut one 2 x 4 x 10′-0″ board into two 4′-5″ lengths for the top and bottom plates.
2. Tack the top and bottom plates together at each end with 6d common nails.
3. Lay out and mark the wall stud locations on the top and bottom plates allowing 3 ½″ for the side wall. See Figure 77.
4. Cut five 2 x 4 x 8′-0″ boards 7′-7 ½″ long for the wall studs.
5. Separate the top and bottom plates.
6. Place the wall studs on layout between the top and bottom plates.
7. Nail the wall studs to the top and bottom plates using two 12d sinkers per stud per plate. See Figure 78.
8. Cut one 2 x 4 x 8′-0″ board 5′-0″ long for the double top plate. This piece will be installed after the wall is stood up.
9. Check the wall frame for square by measuring the diagonals of the wall and adjust as needed.
10. Install a temporary brace to keep the wall square.
11. Stand the wall, aligning the edge of the bottom plate with the outside edge of the floor joist. This will take at least two people.
12. Secure the bottom plate at each end and the center with 12d sinkers through the plywood into the end joist.
13. Align the front wall with the two side walls and fasten the double top plate cut in step 8 to the top plates of the two side walls with 12d sinkers. See Figure 79.

Figure 77
Figure 77 illustrates step 3

Figure 78
Figure 78 illustrates step 7

Figure 79
Figure 79 illustrates step 13

Frame the Gable Roof

This step of the shed construction requires the cutting and assembly of common rafters and a ridge board. If the layout is accurate and the cuts are made carefully, the roof framing will come together nicely. The building of the roof will also require you to toe nail the rafters to the double top plate. This requires practice to avoid splitting the wood.

PROCEDURE

1. Lay out and cut one common rafter pattern using the dimensions provided on the plans. See Figure 80.

Figure 80
Figure 80 illustrates step 1

2. Trace the pattern to lay out and cut 13 more common rafters. Some completed rafters are shown in Figure 81.
3. Mark the rafter layout on the double top plates of the side walls matching the stud layout of the side walls. See Figure 82.
4. Cut the 2 x 6 ridge board to 8′-0″ long, which should be equal to the length of the shed.
5. Lay out both sides of the ridge board matching the rafter layout in step 3.

Figure 81
Completed rafters

Figure 82
Figure 82 illustrates step 3

Figure 83
Figure 83 illustrates step 6

6. Position the ridge board on top of the end walls. See Figure 83.
7. Pull up the pair of opposing common rafters that is at least one layout space in from the two end walls.
8. Place the pair of opposing common rafters on the layout marks on the double top plate.
9. Toe-nail the pair of opposing common rafters to the double top plate with 12d sinkers. See Figure 84.

Figure 84
Figure 84 illustrates step 9

10. Repeat steps 7–9 at the other end of the building.
11. Raise the ridge board between the two pairs of opposing rafters so that the rafters will pinch and hold the ridge board between them.
12. Align the top edge of the rafter flush with the top edge of ridge board and with the previously marked layout on the ridge board. See Figure 85.
13. Nail the ridge board to the rafters with 12d nails. See Figure 86. Refer to the drawing for proper nail placement.

Figure 85
Figure 85 illustrates step 12

Figure 86
Figure 86 illustrates step 13

Figure 87
Figure 87 illustrates step 14

14. Install the remaining rafters on the layout marks and secure them to the ridge board and double top plate with 12d nails. See Figure 87.
15. Align the ridge board even with the end wall and brace it from the front wall up to the ridge board. See Figure 88.

Figure 88
Figure 88 illustrates step 15

alternate project 6 Garden Tool Shed

Garden tools should never be left out in the rain. For this reason, conscientious gardeners often store their tools in outdoor sheds close to where they will be needed. Although they are usually small, garden tool sheds are similar in many ways to much larger structures. Just like larger structures, they have floors, walls, rafters, and a roof. By completing the alternate project outlined in this section, you will learn important fundamentals of building construction. To build the garden shed you must complete nine separate procedures.

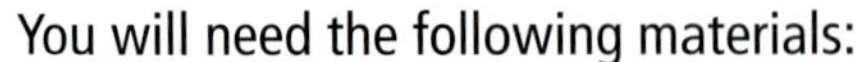

You will need the following materials:

- (6) 2 x 4 x 8′-0″ treated lumber
- (1) ½″ x 4′ x 8′ treated plywood
- (2) 4 x 4 x 8′-0″ treated lumber
- (24) 2 x 3 x 8′-0″ lumber
- (10) 2 x 4 x 8′-0″ lumber
- (4) 2 x 2 x 8′-0″ lumber
- (1) 1 x 3 x 8′-0″ lumber
- (1) sheet ½″ x 4′ x 8′ O.S.B
- (6) sheets ½″ x 4′ x 8′ lap siding material
- (4) Strap hinges
- (1) Hasp
- (1) Cane bolt
- (2 lb) 16d hot dipped galvanized nails
- (2 lb) 16d cement coated (c.c.) sinkers
- (2 lb) 8d c.c. sinkers
- (2 lb) 6d hot dipped galvanized nails
- (6) hurricane ties

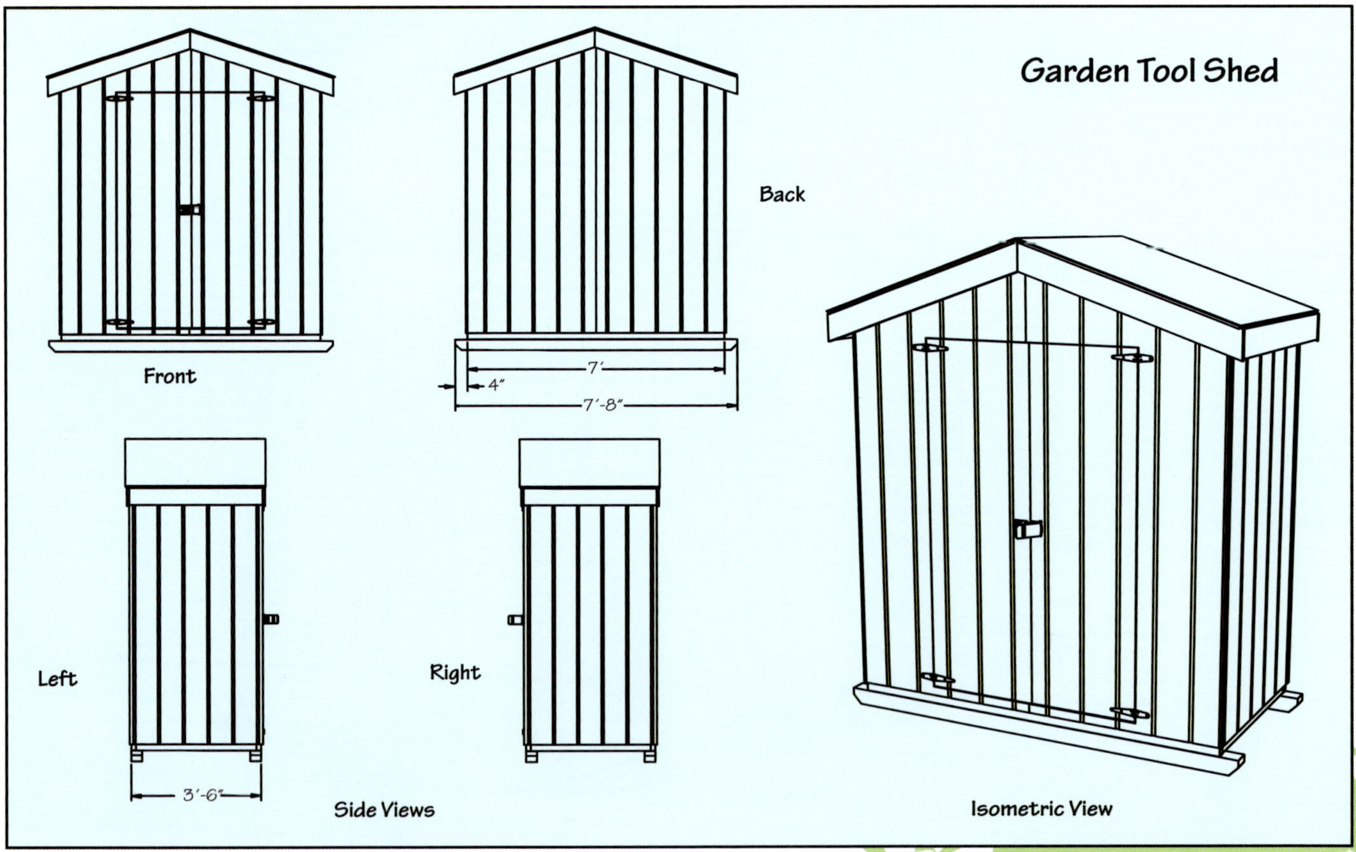

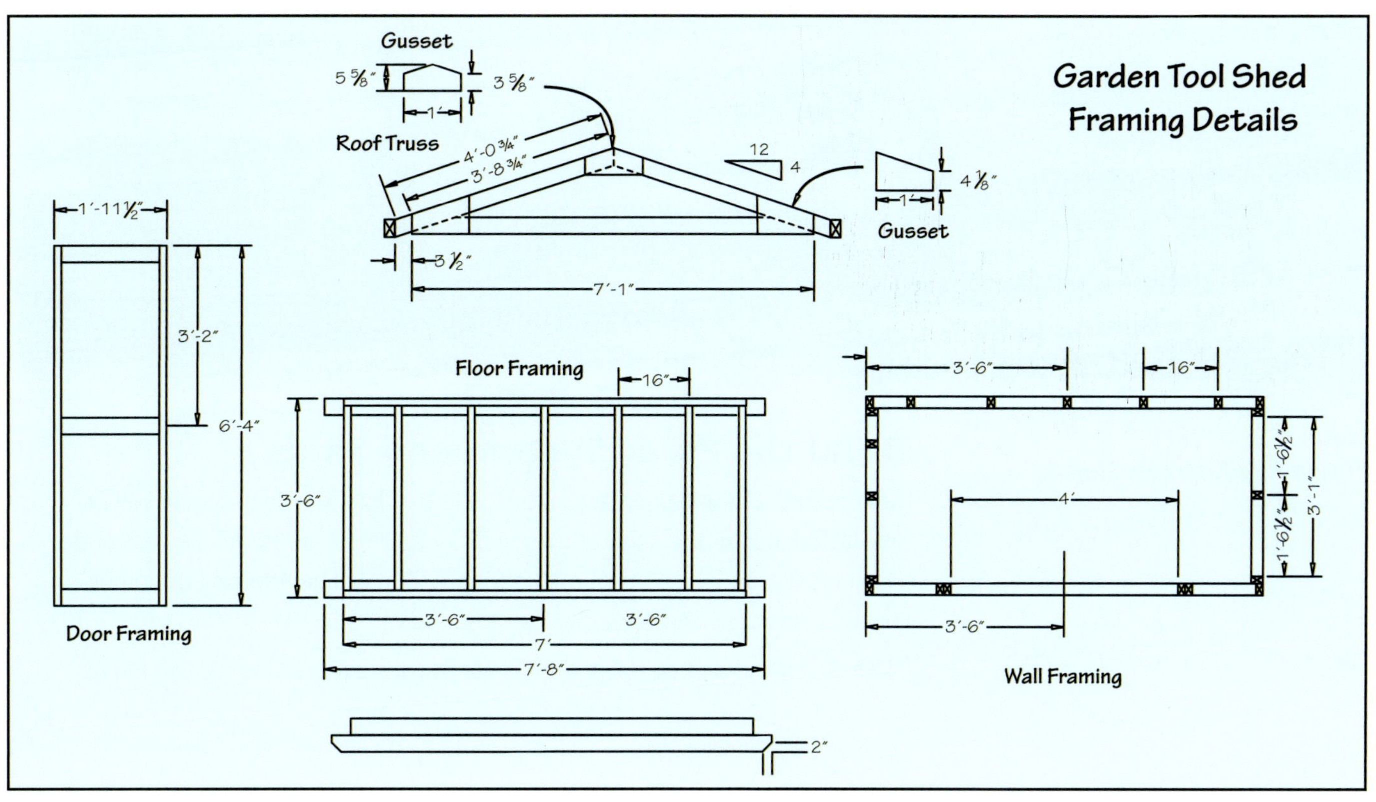

You will need the following tools:

- Chalk box
- Circular saw
- Framing square
- Hammer
- Hand saw
- Pencil
- Electric drill
- Drill index
- Tape measure
- Speed square

Figure 89
Figure 89 illustrates step 1

Build the Floor System and Skids

Construction of the garden shed starts with the floor. The floor system will be assembled atop 4″ x 4″ skids and secured using Simpson/Strong® tie clips and fasteners. The various parts of the flooring system will be attached using hot dipped galvanized 16d nails.

Note: All lumber mentioned in this section is cut from pressure treated lumber.

PROCEDURE

1. Cut two 4 x 4 x 7′-8″ skids and bevel both ends. See Figure 89.
2. Cut two 2 x 4 x 7′-0″ rim joists and seven 2 x 4 x 3′-3″ floor joists.
3. Mark the location of the floor joists on the rim joists with a 16″ center layout following the framing plan.
4. Assemble the framing for the floor system with 16d hot dipped galvanized nails.
5. Center the floor system framing on the 4 x 4 skids, aligning the outside face of the rim joists with the outside face of the skids. See Figure 90.

Figure 90
Figure 90 illustrates step 5

6. Secure the floor system framing to the skids with six hurricane ties.
7. Lay out and cut the treated plywood to 7′-0″ long and 3′-6″ wide. See Figure 91.

Check the plywood for square...

Figure 91
Figure 91 illustrates step 7

8. Fasten the plywood onto the floor joist system with 6d hot dipped galvanized nails, aligning the edges of the plywood to the edges of the floor system. See Figure 92.

Figure 92
Figure 92 illustrates step 8

Build the Rear Wall

For the rear wall you will cut two pieces of 2 x 3 x 7′-0″ lumber for use as plates and seven 6′-3″ studs. The plates will be fastened to the studs with sinkers. Then you will cut two sheets of siding and fasten them to the wall with 6d hot dipped galvanized nails. All of the walls—rear, front, and sides—will be framed and sided flat on the work surface and then raised on the floor system.

PROCEDURE

1. Cut two 2 x 3 x 7′-0″ boards for the top and bottom plates for the rear wall.
2. Lay out the location of the studs on the plates using 16″ o.c. spacing laid out from the center of the plate.
3. Cut seven 2 x 3 studs 6′-3″ long for the rear wall.
4. Fasten the plates to the studs with 16d c.c. sinkers.
5. Cut a sheet of siding to a width of 3′-6″ measuring from the tongue edge of the sheet.
6. Align the cut edge of the sheet with the ends of the top and bottom plates and leave a 2″ overhang past the bottom plate. See Figure 93.
7. Fasten the sheet to the face of the wall, with 6d hot dipped galvanized nails.
8. Cut a second sheet for the remaining portion of the rear wall to width and fasten to the wall in the same manner as done in steps 6 and 7.
9. Measure 6′-10″ from the bottom of the siding along each cut end of the wall and place a mark. See Figure 94.
10. Mark a line from the marks made in step 9 through center point of the top edge of the siding. See Figure 95.
11. Cut along this line with the circular saw.

Figure 93
Figure 93 illustrates step 6

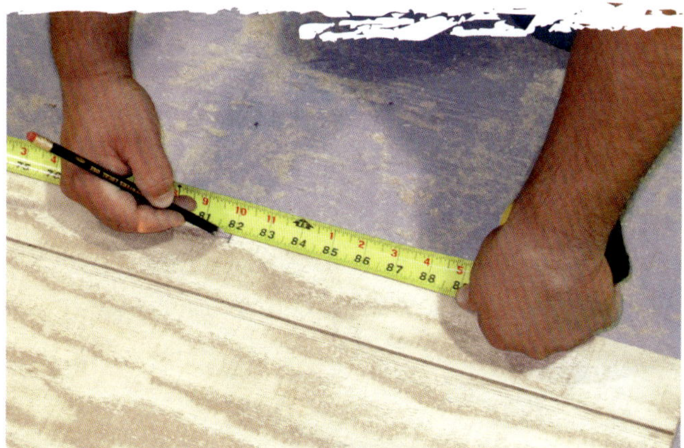

Figure 94
Figure 94 illustrates step 9

Figure 95
Figure 95 illustrates step 10

Build the Front Wall

The process of building the front wall is more complicated than that of building the rear wall since you must leave an opening for the door. You must very carefully measure the door opening and transfer these measurements to the siding. You will use a chalk line to establish the cutting line on the siding and then make the cut using a circular saw. Some of the cut pieces will be set aside for use in making the doors.

PROCEDURE

1. Cut two 2 x 3 x 7′-0″ boards for the top and bottom plates for the front wall.
2. Lay out the plates centering a 4′-0″ door opening with double 2 x 3s at each side of the door opening and a 2 x 3 stud at each end. See Figure 96.

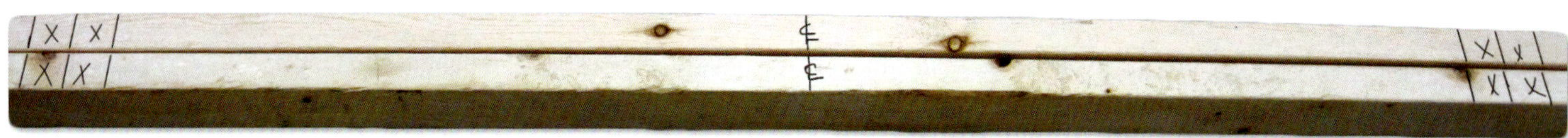

Figure 96
Figure 96 illustrates step 2

3. Cut six studs 6′-3″ long for the front wall.
4. Fasten the plates to the studs with 16d c.c. sinkers.
5. Cut a sheet of siding to a width of 3′-6″ measuring from the tongue edge of the sheet.
6. Fasten the cut sheet to the face of the wall, aligning the cut edge of the sheet with the ends of the top and bottom plates, and a 2″ overhang past the bottom plate, with 6d hot dipped galvanized nails. Do not fasten the siding to the bottom plate where the door will be located.
7. Measure from the tongue edge of the sheet to the door stud at the top and bottom of the opening and transfer the dimensions to the face of the sheet.
8. Snap a chalk line through the marks made in step 7 to lay out the cut line for the edge of the door in this sheet.
9. Measure the dimension from the bottom of the siding to the bottom edge of the top plate and transfer that dimension to the cut line laid out in step 8.
10. Measure from the chalk line made in step 8 to the edge of the exposed door stud at the top and bottom plates and note these dimensions for later use.
11. Cut the second sheet for the front wall to width and fasten to the wall in the same manner as done in step 6.
12. Lay out the second door cut line using the dimensions noted in step 10.
13. Measure along the cut line in step 12 the dimension measured in step 9 to locate the top of the door.
14. Snap a line through the marks made in steps 9 and 13. See Figure 97.

Figure 97
Figure 97 illustrates step 14

Figure 98
Figure 98 illustrates step 16

15. Measure up 6′-10″ from the bottom of the siding along each cut end of the wall and place a mark.
16. Mark cut lines from the marks made in step 15 through the center point of the top edge of the siding. See Figure 98.
17. Check the circular saw to a depth that will cut through the siding, but no deeper.
18. Cut the door panels by making a saw cut through the overlap of both sheets of siding, from the bottom of the sheets to the top door cut line, both door side cut lines, and across the top cut line. See Figure 99.
19. Cut the lines made in step 16 with the circular saw.
20. Reserve and label these panels and mark the top plate with matching marks for later use.

Figure 99
Figure 99 illustrates step 18

Build the End Walls

To build the end walls you will cut four pieces of 2 x 3 x 3′-1″ lumber for use as plates and six pieces of 2 x 3 x 6′-3″ lumber for use as studs. The plates will be fastened to the studs using sinkers. Then you will cut two pieces of siding and fasten them to the wall using 6d hot dipped galvanized nails.

PROCEDURE

1. Cut four 2 x 3 x 3′-1″ boards to use as the top and bottom plates for the end walls.
2. Cut six 2 x 3 x 6′-3″ studs for the end walls.
3. Position the studs between the top and bottom plates with one at each end and one in the center and fasten with 16d c.c. sinkers.
4. Cut the two remaining sheets of siding to 3′-6″ wide by 6′-8″ high for the two end walls.
5. Center the siding panels on the wall frames side to side while aligning the top edges with the top plates and fasten with 6d hot dipped galvanized nails. See Figure 100. Be certain that the distance from the edge of the siding panel to the side of the end studs is equal at the top, center, and bottom of the stud.

Figure 100
Figure 100 illustrates step 5

Attach the Walls to the Floor System

During this procedure, the garden shed structure will begin to take shape as you attach the walls to the floor system. This is accomplished primarily by aligning each wall in its proper position and fastening the bottom plate to the floor joist with hot dipped galvanized nails. The wall panels are fastened to one another with sinkers.

PROCEDURE

1. Stand the rear wall on the floor system, aligning the ends of the wall with the ends of the floor system and temporarily support it. See Figure 101.
2. Press the bottom edge of the wall panel against the rim joist and fasten the siding to the rim joist with 6d hot dipped galvanized nails. See Figure 102.

Figure 101
Figure 101 illustrates step 1

Figure 102
Figure 102 illustrates step 2

Figure 103
Figure 103 illustrates step 3

3. Nail the bottom plate through the treated plywood into the floor joist with 16d hot dipped galvanized nails. See Figure 103.
4. Stand the end wall on the floor system while continuing to temporarily support the rear wall. Position the end wall so the bottom plate is against the end stud of the rear wall. Fasten the bottom of the siding to the joist in a similar manner as used in step 2. See Figure 104.

Figure 104
Figure 104 illustrates step 4

Figure 105
Figure 105 illustrates step 5

5. Nail 16d c.c. sinkers through the end stud of the end wall into the end stud of the rear wall while maintaining contact of the siding of the end wall with the end stud of the rear wall. See Figure 105.
6. Nail the bottom plate to the floor joist with 16d hot dipped galvanized nails.
7. Nail the edge of the overlapping siding to the end stud of the rear wall with 6d hot dipped galvanized nails. See Figure 106.
8. Repeat steps 4–7 to attach the second end wall.
9. Stand the front wall on the floor system with the ends of the wall flush with the ends of the floor system.
10. Fasten the bottom of the siding to the rim joist in a similar manner used in step 2.
11. Nail 16d c.c. sinkers through the end studs of the end walls into the end studs of the front wall while maintaining contact of the siding of the end wall with the end studs of the front wall.
12. Nail the bottom plate to the floor joist with 16d hot dipped galvanized nails, making sure not to drive nails through the plate at the door opening.
13. Cut and remove the bottom plate of the front wall at the door opening with the hand saw.

Figure 106
Figure 106 illustrates step 7

Cut and Assemble the Roof Trusses

Now you will begin the process of constructing the garden shed roof by cutting and assembling the trusses. These are the triangular members that support the roof. The upper part of the truss that actually holds up the roof is called a top chord. First you will cut a pattern and then use it to cut five additional top chords and four fly rafters. Then you will cut the bottom chords and the gussets that will link the bottom chord to the top chords. The various parts will be nailed together using 6d hot-dipped galvanized nails.

Figure 107
Figure 107 illustrates step 1

PROCEDURE

1. Lay out and cut a top chord pattern for the truss using the information shown in the plans. See Figure 107.
2. Trace and cut nine more top chords using the pattern cut in step 1. Set four of the top chords aside to be used as fly rafters in the next procedure.
3. Lay out and cut a pattern for the bottom chord of the truss using the information shown in the plans. See Figure 108.
4. Trace and cut two more bottom chords using the pattern cut in step 3.

Figure 108
Figure 108 illustrates step 3

Figure 109
Figure 109 illustrates step 5

5. Cut one pattern gusset for the joint at the ridge of the truss and one pattern gusset for the joint at the bottom chord of the truss. See Figure 109.
6. Trace and cut two additional ridge gussets and five additional bottom chord gussets using the patterns made in step 5. Gussets will placed on one side only on each truss.
7. On a flat surface, position two top chords and one bottom chord and nail them together.
8. Fasten the ridge gusset and bottom chord gussets to the truss using 6d hot dipped galvanized nails. See the completed truss in Figure 110.
9. Repeat steps 7 and 8 to construct the remaining 2 trusses.

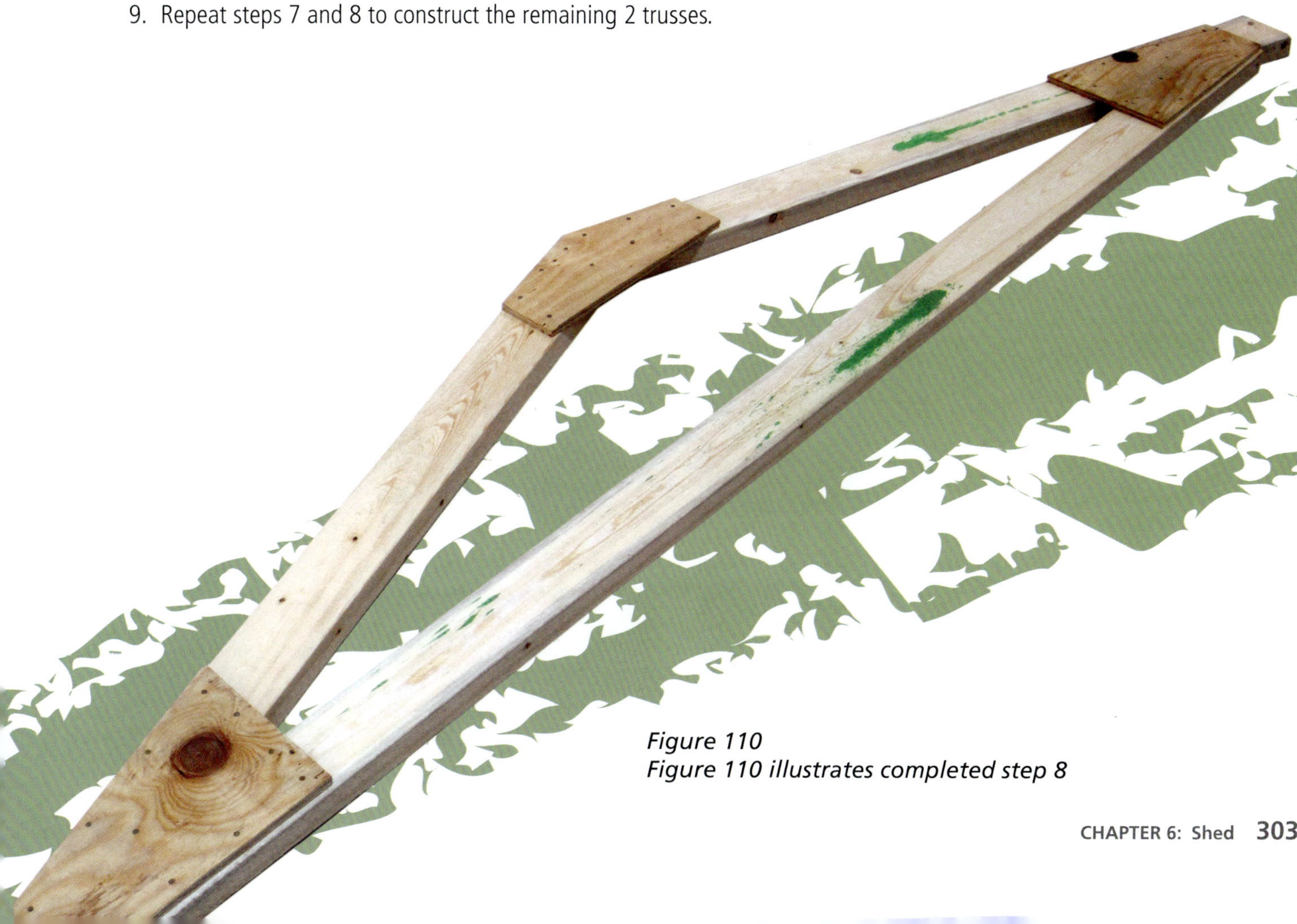

Figure 110
Figure 110 illustrates completed step 8

Set the Trusses and Fly Rafters

Once you have built the trusses, they will be set in place on top of the walls of the garden shed and then fastened in place using nails. You will begin this process at the front wall and then repeat it at the rear wall. Then you will place and attach the sub fascia or backer boards to the fly rafters.

PROCEDURE

1. Center a truss on the top of the front wall, with the gussets facing inward and the top and bottom chords firmly against the siding. Fasten the truss through the siding with 6d hot dipped galvanized nails and toe-nail the bottom chord to the top plate with 8d c.c. sinkers.
2. Repeat step 1 for the truss on the rear wall.
3. Cut a temporary spreader 3′-10″ long and lay it out to the dimensions shown in the plans.
4. Temporarily attach the spreader to the front and rear trusses at the ridge with one end flush to the siding on the rear truss.
5. Center the third truss on the end walls and use a straightedge placed against the tails of the front and rear trusses to align the tail of the center truss. Tack the third truss to the temporary spreader. See Figure 111.
6. Toe-nail the truss to the top plates with 8d c.c. sinkers.
7. Temporarily nail through the spreader at the ridge.
8. Cut two 2 x 4 x 3′-10″ sub fascia boards.

Figure 111
Figure 111 illustrates step 5

Figure 112
Figure 112 illustrates step 9

Figure 113
Figure 113 illustrates step 10

9. Fasten sub fascia boards to the top chord tails on each side of the truss with 16d c.c. sinkers. See Figure 112.
10. Position one fly rafter against the back side of the sub fascia and against the underside of the temporary spreader. See Figure 113.
11. Fasten the fly rafter through the sub fascia with 16d c.c. sinkers. Temporarily fasten the fly rafter to the spreader at the top with 8d c.c. sinkers.
12. Position the second fly rafter against the backside of the sub fascia on the other side of the shed and against the plumb cut of the first fly rafter. See Figure 114.
13. Fasten the fly rafters with 16d c.c. sinkers through the sub fascia. Then, toe-nail the two fly rafters together at the ridge with 8d c.c. sinkers.

Figure 114
Figure 114 illustrates step 12

Cut and Install the Roof Sheathing

The roof sheathing of the garden shed is made of oriented strand fiber board (OSB). You will cut two pieces of OSB to size and then nail them to the trusses with sinkers. Make sure the OSB is properly aligned.

PROCEDURE

1. Measure the distance from the face of the sub fascia to the ridge line of the truss along the top edge of the truss.
2. Cut two pieces of ½″ OSB to 3′-10″ long and to the width measured in step 1.
3. Align the bottom edge of the sheet on the side opposite the spreader with the vertical face of the sub fascia and flush to the outside face of the siding on the rear of the building. When in position, fasten to the top chord tails with 6d c.c. sinkers.
4. Nail the OSB to the remaining two trusses with 6d c.c. sinkers.
5. Align the fly rafter with the edge of the OSB and fasten with 6d c.c. sinkers. See Figure 115.
6. Remove the temporary spreader from the trusses.
7. Repeat steps 3–5 to install the OSB on the second side of the roof.

Figure 115
Figure 115 illustrates step 5

Make and Install the Doors

The final procedure in construction of the garden shed involves making and installing the doors. The door panels were cut, marked, and set aside during an earlier procedure. The frame will be constructed using pieces cut from 2 x 4 and 2 x 2 lumber. The door frame will be constructed using 2 x 2s for the vertical members and 2 x 4s for the horizontal members. Once the doors are installed, the garden shed project is complete.

PROCEDURE

1. Select one labeled door panel reserved earlier, measure its length and width, and note these dimensions. Be certain to make the doors so they can be installed in their proper location as indicated by the earlier labeling and marking.
2. Make the door frame width ⅛″ narrower than the width of the panel and 2 ¼″ shorter than the length.
3. Fasten the 2 x 2s to the 2 x 4s with 16d c.c. sinkers. See Figure 116.
4. Place the door panel on the assembled frame aligning the top edge of the panel with the top edge of the frame and the hinge side of the panel with the hinge side of the frame. See Figure 117.
5. Fasten the panel to the frame with 6d hot dipped galvanized nails 4″ o.c.
6. Repeat steps 1–5 for the second door.

Figure 116
Figure 116 illustrates step 3

Place the door panel on the assembled frame...

Figure 117
Figure 117 illustrates step 4

Figure 118
Figure 118 illustrates step 7

7. Attach two strap hinges to each door with the screws provided with the hinges. See Figure 118.
8. Place a ¼″ shim on the shed floor at the door opening and position the door in the door opening.
9. Adjust the door to maintain a ⅛″ margin on the hinge side and a ⅛″ margin at the top. See Figure 119.

Figure 119
Figure 119 illustrates step 9

Figure 120
Figure 120 illustrates step 10

Figure 121
Figure 121 illustrates a completed step 12

10. Secure the hinge through the siding into the double stud of the door opening with the screws provided by the manufacturer. See Figure 120.
11. Repeat steps 7–10 to install the second door.
12. Center and install the hasp. See Figure 121 for the completed installation.
13. Drill a hole for the cane bolt and install. See Figure 122.

Figure 122
Figure 122 illustrates a completed step 13

Drill a hole for the cane bolt...

alternate project 7 Wishing Well

As an alternate project you may decide to build a wishing well. Some homeowners use wishing wells as ornamental garden structures. Building a wishing well requires some of the same carpentry skills and techniques needed for construction of any structure. Only four procedures are required to build the wishing well.

You will need the following materials:

- (5) 1 x 4 x 8′-0″ rough sawn cedar or treated lumber
- (2) 1 x 8 x 8′-0″ rough sawn cedar or treated lumber
- 1″ diameter x 30″ wood dowel
- ¼″ diameter x 6″ wood dowel
- 13 lineal feet 1 ½″ galvanized strap
- 2 lb ¾″ zinc-plated pan-head screws
- 1 lb 1 ¼″ deck screws

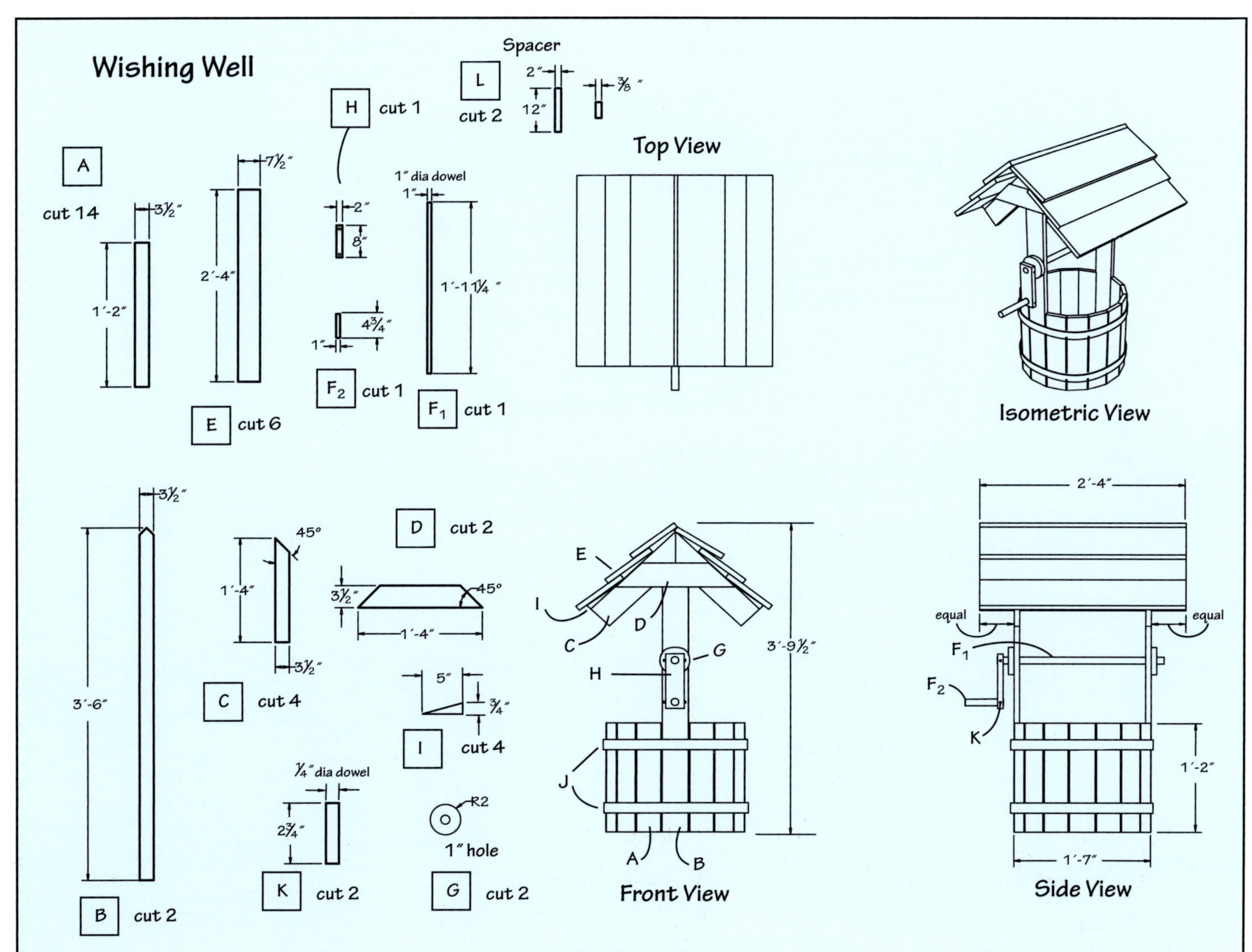

You will need the following tools:

- (2) Adjustable bar clamps
- Aviation snips
- Circular saw
- Combination square with center head
- Compass
- Countersink bit
- Dovetail saw
- Drill index
- ⅜″ drill
- Extension cord
- Hammer
- Handsaw
- Miter saw
- Pencil
- Phillips head screwdriver
- Retractable tape measure
- Saber saw
- Safety glasses, hard hat, work shoes, and other PPE
- Sand paper
- Screw gun
- 1″ spade bit
- ¼″ and ¾″ wood chisel
- Wood rasp
- (2) Sawhorses

Cutting the Pieces

The first procedure in the construction of your wishing well is to cut the pieces. You will have to create a cut list from the materials list and the drawing and then use a circular saw or hand saw to cut most of the parts.

Assembling of the Palings

In the previous procedure you cut (14) 1 x 4 x 1′-4″ cedar palings and two 1 x 4 x 3′-6″ extended cedar palings. In this procedure you will use these palings to form the outer wall of the wishing well. The palings will be attached to a galvanized strap using deck screws and then rolled into a circle.

PROCEDURE

1. Mark a line across the face of each paling 1 ½″ from each end.
2. Place one extended paling followed by seven palings. Place the other extended paling followed by the remaining seven palings.
3. Butt the ends of the palings against a straight edge to keep them in a straight line. See Figure 123.

Figure 123
Figure 123 illustrates step 3

Figure 124
Figure 124 illustrates step 4

4. Align the 1 ½" perforated galvanized straps with the marks made in step 1. See Figure 124. Place the bottom strap above the line and the top strap below the line.
5. Maintain a ⅜" gap between each paling using the spacer.
6. Attach the perforated galvanized straps to each paling using two ¾" deck screws per strap per paling while maintaining the necessary ⅜" gap. See Figure 125.
7. Turn the assembly over and roll it into a circle.
8. Stand the assembly on edge and overlap the galvanized straps by the width of at least one paling board.
9. Cut off the excess galvanized strap.
10. Screw both overlapping straps to the paling using two ¾" pan-head screws. See Figure 126.

Figure 125
Figure 125 illustrates step 6

Figure 126
Figure 126 illustrates step 10

Assembling of the Roof

To construct the roof of the wishing well you will build two rafter assemblies by attaching rafters to a horizontal member called a collar tie. The completed assemblies will be attached to extended palings with deck screws.

PROCEDURE

1. On a flat surface, butt two opposing rafters together joining the 45° angles. See Figure 127.

Figure 127
Figure 127 illustrates step 1

2. Place a collar tie on top of the adjoined rafters. See Figure 128.
3. Attach the collar tie using two 1 ¼" deck screws per connection to each rafter.
4. Repeat steps 1–3 for the opposing rafter assembly.
5. Clamp one of the rafter assemblies to an extended paling, matching the angles at the top of the paling with the underside of the rafter assembly. Collar ties are located on the outside.

Figure 128
Figure 128 illustrates step 2

6. Attach the rafter assembly to the extended paling using two 1 ¼″ deck screws. See Figure 129.
7. Repeat steps 5 and 6 for the opposing rafter assembly. See Figure 130.

Figure 129
Figure 129 illustrates step 6

Figure 130
Figure 130 illustrates step 7

Figure 131
Figure 131 illustrates step 8

8. Place one ¾″ x 5″ wedge along the bottom of each rafter assembly. See Figure 131. Attach the wedge using one 1 ¼″ deck screw per wedge.
9. Starting at the end of the rafter assembly, attach the first roof board using 1 ¼″ deck screws, maintaining a ¾″ overhang from the end of the rafters and an equal overhang at the extended palings. See Figure 132.
10. Overlap and attach the two remaining roof boards to the rafters with 1 ¼″ deck screws, while ensuring equal exposure of the roof boards.
11. Repeat steps 9 and 10 to complete the opposite side of the roof.

Figure 132
Figure 132 illustrates step 9

(A)

Assembling of the Handle and the Spindle

A handle and spindle help give the wishing well its characteristic appearance. To assemble the handle and spindle you will drill a 1″ hole in each of the extended palings using a spade bit. Then you will cut an arm, handle, and spindle to length and attach them using deck screws.

PROCEDURE

1. Measuring from the floor, make a mark on each extended paling at 30″.
2. Center the mark made in step 1 on the extended paling.
3. Using the 1″ spade bit, drill a hole in each extended paling at the intersecting marks made in steps 1 and 2.
4. To create a washer, draw a 4″ diameter circle on a piece of 1 x 8 material, as shown in Figure 133(A), and bore a center hole with the 1″ spade bit. See Figures 133(B) and (C).

Figure 133
Figure 133 illustrates step 4

(B)

(C)

5. Cut the 4″ diameter circle with the saber saw. See Figure 134(A). Smooth the edges with sandpaper. See Figure 134(B).

(A)

(B)

Figure 134
Figure 134 illustrates step 5

Cut the 4″ diameter circle with the saber saw...

(a) (b)

Figure 135
Figure 135 illustrates step 6

6. Drill a pilot hole, see Figure 135a, and countersink in the edge of each washer, shown in Figure 135b.
7. Lay out and mark the arm of the spindle and handle assembly per the dimensions on the diagram.
8. Cut the spindle and the handle to length as per the dimensions on the diagram.
9. Insert the handle into the arm.
10. Drill a ¼″ hole completely through the arm and handle.
11. Repeat steps 8 and 9 to install the spindle into the arm.
12. Cut the ¼″ x 6″ wood dowel into two 2 ¾″ dowels with the dovetail saw.
13. Attach the handle and spindle to the arm with a 2 ¾″ x ¼″ dowel. Be sure to use glue on the dowels. See Figure 136.
14. Place a 4″ washer on the spindle, butting against the arm. See Figure 137.

Figure 136
Figure 136 illustrates step 13

Figure 137
Figure 137 illustrates step 14

15. Insert the spindle assembly into the holes on the extended palings.
16. Place the remaining 4″ washer on the end of the spindle
17. Attach both 4″ washers to the spindle with 2″ deck screw. See Figure 138.

Figure 138
Figure 138 illustrates step 17

Student Name: ______________________________ Date: ______________

Shed with Gable Roof Project Evaluation

PROCEDURE	CRITERIA AS SPECIFIED BY THE PROJECT	POSSIBLE POINTS	SCORE
Floor System	Joist lay out	5	
	Rim joists attachment	5	
	Plywood installation	5	
	Parallel sides equal	5	
	Floor length	5	
	Floor width	5	
	Floor system square	5	
	Fastener installation	5	
	Subtotal	*40*	
Wall System	Stud lay out	5	
	Rough opening for door	5	
	Door opening frame	5	
	Rough opening for window	5	
	Window opening frame	5	
	Header location	5	
	Headers built correctly	5	
	Walls square	5	
	Walls flush at corners	5	
	Double plate installation	5	
	Fastener installation	5	
	Subtotal	*55*	
Roof System	Rafter lay out	5	
	Rafter pitch correct	5	
	Plumb cuts	5	
	Birdsmouth cuts	5	
	Overhang correct	5	
	Ridge board installation	5	
	Gable stud installation	5	
	Temporary bracing installation	5	
	Fastener installation	5	
	Subtotal	*45*	
Overall Dimensions	Length correct	5	
	Width correct	5	
	Wall height correct	5	
	Subtotal	*15*	
General	Proper tool handling	5	
	Followed direction	5	
	Cleaned up	5	
	Safe work practices	5	
	Subtotal	*20*	
		Student score:	

Total Possible Points = 175

Suggested minimum acceptable score: 123 points or ______

Student's Signature: ______________________ Teacher's Signature: ______________________

Student Name: ______________________________ Date: ______________

Garden Tool Shed Project Evaluation

PROCEDURE	CRITERIA AS SPECIFIED BY THE PROJECT	POSSIBLE POINTS	SCORE
Floor System	Joist lay out	5	
	Rim joists attachment	5	
	Plywood installation	5	
	Parallel sides equal	5	
	Floor system square	5	
	Fastener installation	5	
	Subtotal	*30*	
Wall System	Stud lay out	5	
	Rough opening for door	5	
	Walls square	5	
	Walls flush at corners	5	
	Double plate installation	5	
	Siding installation	5	
	Fastener installation	5	
	Subtotal	*35*	
Roof System	Top chord lay out	5	
	Bottom chord lay out	5	
	Truss built correctly	5	
	Rafter pitch correct	5	
	Overhang correct	5	
	Sheathing installation	5	
	Fastener installation	5	
	Subtotal	*35*	
Overall Dimensions	Length correct	5	
	Width correct	5	
	Wall height correct	5	
	Subtotal	*15*	
General	Proper tool handling	5	
	Followed direction	5	
	Cleaned up	5	
	Safe work practices	5	
	Subtotal	*20*	
		Student score:	

Total Possible Points = 135

Suggested minimum acceptable score: 95 points or ______

Student's Signature: ____________________ Teacher's Signature: ____________________

Student Name:_______________________ Date: _______________

Wishing Well Project Evaluation

PROCEDURE	CRITERIA AS SPECIFIED BY THE PROJECT	POSSIBLE POINTS	SCORE
Cutting	Paling (14)	35	
	Extended paling (2)	5	
	Rafters (2)	5	
	Collar ties (4)	10	
	Roof boards (6)	15	
	Dowel	5	
	Spindle and handle	5	
	Wedges (4)	5	
	Galvanized straps (2)	5	
	Plywood	5	
	Subtotal	*95*	
Assembly - Palings	Palings	5	
	Galvanized strap	5	
	Fastener installation	5	
	Subtotal	*15*	
Assembly - Roof	Collar tie	10	
	Rafter	5	
	Fastener installation	5	
	Subtotal	*20*	
Assembly Handle & Spindle	Handle and spindle	5	
	Fastener installation	5	
	Subtotal	*10*	
Overall	Height correct	5	
	Joints tight	5	
	No rough edges	5	
	Edges flush	5	
	No splits or shiners	5	
	Subtotal	*25*	
General	Proper tool handling	5	
	Followed direction	5	
	Cleaned up	5	
	Safe work practices	5	
	Subtotal	*20*	
		Student score:	

Total Possible Points = 185

Suggested minimum acceptable score: 130 points or ______

Student's Signature: _______________________ Teacher's Signature: _______________________

CHAPTER 7

PLAYHOUSE

CONTENTS

Introduction

This chapter provides the procedures for a single but challenging project—a playhouse. Even though this project is designed to provide pleasure for children, its construction is far from child's play. In some ways, building this playhouse is a lot like building a miniature house. Consequently, this project will provide you with many opportunities to practice the carpentry skills that are vitally important to your growth as a professional carpenter.

What's New?

Building Paper Specially treated paper used in wall and roof framing to help protect sheathing from water is called building paper. See Figure 1.

Figure 1
(A) Building paper for wall
(B) Building paper for roof

(A)

(B)

Drip Edge L-shaped metal, called drip edge, is used around the edges of a roof to help water drip clear of underlying construction. See Figure 2.

Post Base A post base is a metal fastener used to secure a post to a floor. See Figure 3.

...to help water drip clear...

Figure 2
Drip edge

Figure 3
Post base

Figure 4
Post cap

Post Cap A metal fastener used to secure a beam to the top of a post is called a post cap. See Figure 4.

Shingles The roofing material placed over building paper, sheathing, and framing to protect the structure from water is called shingles. See Figure 5.

Figure 5
Shingles

Playhouse

Building a full size playhouse will require you to use many of the carpentry skills you have learned so far and even test some of your skills against new applications. For example, in this project, you may build a complete porch with a railing. Since this playhouse is like a tiny house, it would make a very special gift for anyone who has young children. And because it is made of weather resistant materials such as cedar, it should last a very long time if properly maintained. The construction of the playhouse will require seven procedures.

You will need the following materials:

Floor Framing:

- (2) 2 x 6 x 8′-0″ treated lumber for rim joists
- (7) 2 x 6 x 8′-0″ treated lumber for floor joists
- (3) 4 x 4 x 8′-0″ treated lumber for skids
- (14) 3″ x 3″ x 3″, 16 gauge galvanized steel for joist clip angles
- (2) ¾″ x 4′-0″ x 8′-0″ exterior grade plywood for floor sheathing
- (6) 1 x 6 x 10′-0″ treated lumber for decking

Wall Framing:

- (10) 2 x 4 x 8′-0″ treated lumber for plates and studs
- (20) 2 x 4 x 10′-0″ treated lumber for plates and studs
- (2) 2 x 6 x 12′-0″ treated lumber for headers
- (3) pieces ½″ x 5″ x 3′-0″ sheathing for header spacers
- (7) sheets ½″ x 4′-0″ x 8′-0″ exterior grade plywood for sheathing
- (7) sheets ¼″ x 4′-0″ x 8′-0″ paneling for interior paneling (optional)

Porch Framing:

- (1) 4 x 4 x 10′-0″ cedar or treated lumber for post
- (1) 4 x 6 x 10′-0″ cedar or treated lumber for beam or built up 2x6 header
- (2) post bases 4 x 4
- (2) post beam caps 4 x 4

Roof Framing:

- (18) 2 x 4 x 8′-0″ lumber for rafters, lookouts, and collar ties
- (1) 2 x 6 x 10′-0″ lumber for ridge board

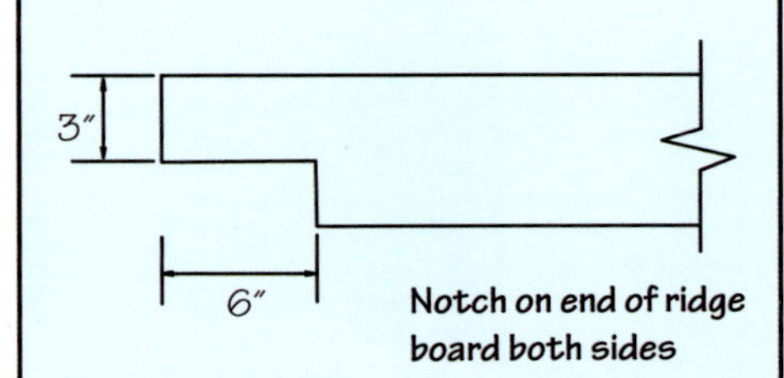

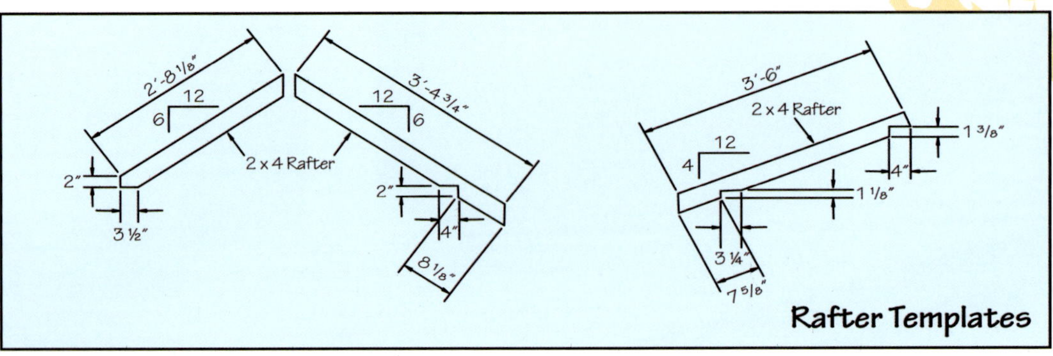

Roofing:

- (4) ½″ x 4′-0″ x 8′-0″ plywood for roof sheathing
- (100 sq. ft.) 15 lb building paper
- 4 bundles of 240 lb. 12″ x 3′-0″, 3 tab self sealing asphalt shingles
- (2) 10′-0″ pieces metal drip edge
- (2) 10′-0″ pieces of metal gutter flashing

Exterior Finishes:

- (248 lin. ft.) 8″ lap siding (optional)
- (4) 1″ x 6″ x 10′-0″ cedar for fascia
- (2) 1″ x 6″ x 8′-0″ cedar for deck trim (optional)

Railings: (optional)

- (1) 2″ x 4″ x 8′-0″ cedar for top rail
- (8) 2″ x 2″ x 2′-6″ cedar for balusters
- (1) 2″ x 2″ x 8′-0″ cedar for bottom rail or nailer

Fasteners:

- (3 ½ lbs.) 16d galvanized nails
- (2 lbs.) 16d common nails
- (1 lb.) 12d common nails
- (5 lbs.) 8d galvanized nails
- (½ lb.) 6d galvanized finish nails (optional)
- (2 lbs.) 6d nails
- (2 lbs.) 5d siding nails (optional)
- (60) 1 ½″ joist hanger nails
- (1 lb.) ⅞″ galvanized roofing nails
- (120) 2″ deck screws
- (36) 2 ½″ deck screws
- (4) tubes construction adhesive (optional)
- (2) tubes silicon latex caulk (optional)

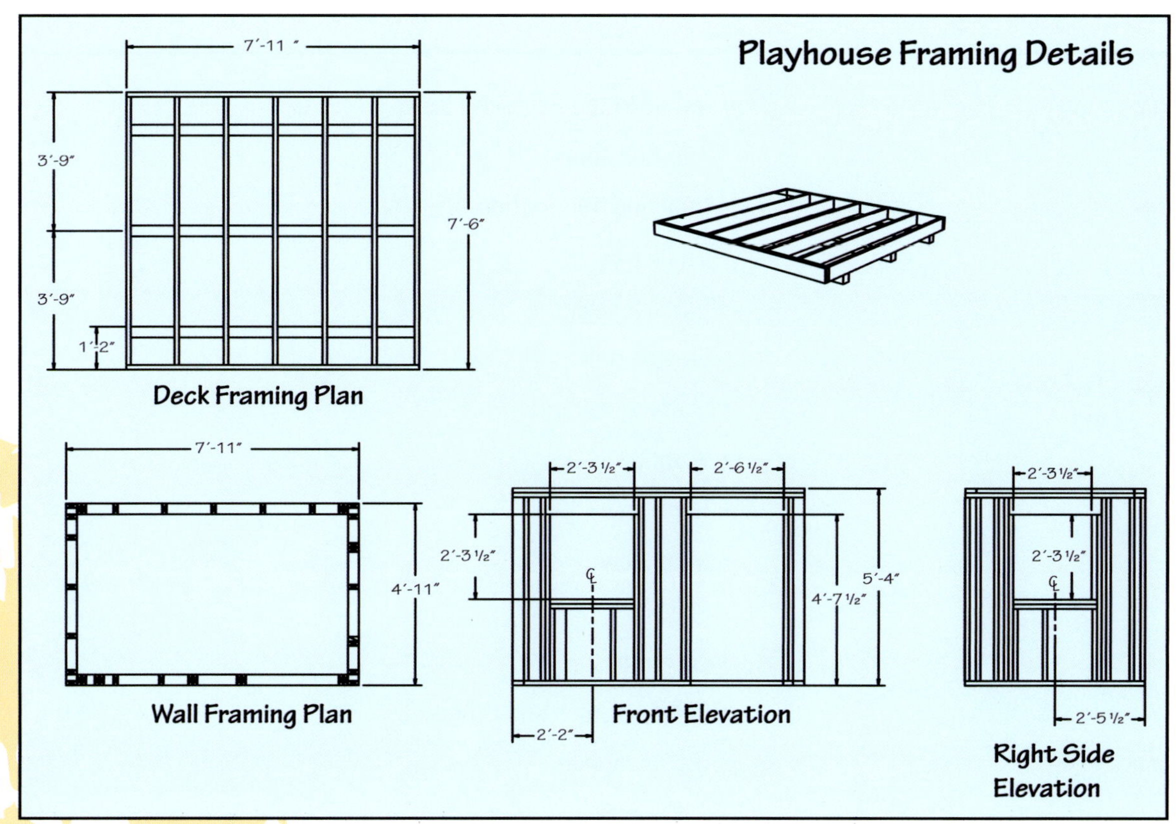

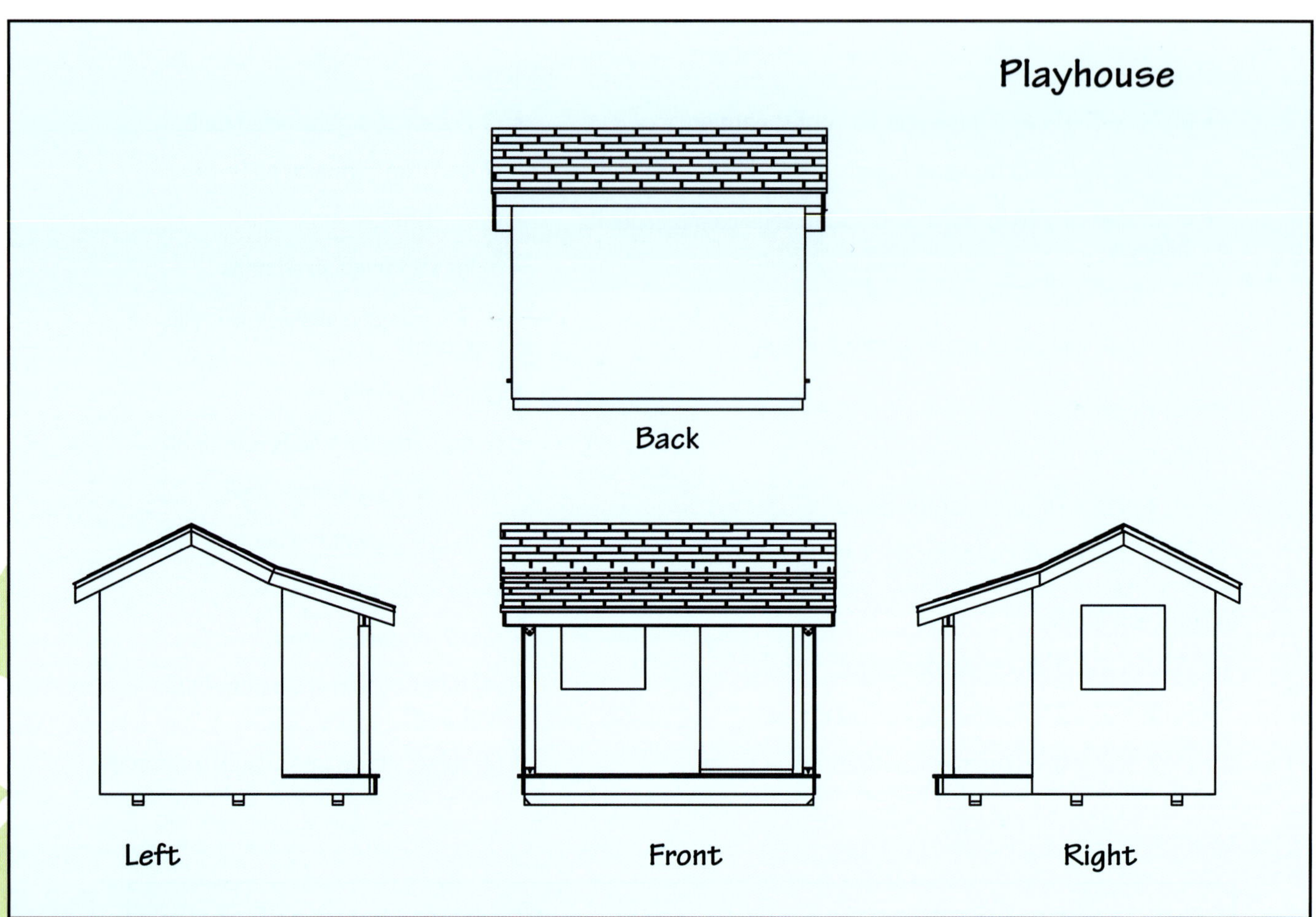

You will need the following tools:

- Aviation snips
- Caulking gun (optional)
- Chalk box
- Circular saw
- Electric drill
- Framing square
- Hammer
- Hand saw
- Miter saw
- Pencil
- Tape measure
- Safety glasses and other PPE
- (2) Sawhorses
- Screw gun
- Speed square
- Utility knife

Build the Floor Frame

Construction of the playhouse begins with building the floor frame. You will start by cutting three 4 × 4 treated timbers to a length of 7′-11″ for use as skids. The skids will make it possible to move the playhouse when necessary. Next you will cut the rim joists and floor joists and assemble the floor frame using 16d galvanized nails. Plywood floor sheathing and decking boards will be installed on top of the frame.

PROCEDURE

1. Cut three 4 x 4 treated timbers 7′-11″ long for the skids and bevel the ends 45°.
2. Cut two 2 x 6 x 7′-11″ rim joists and seven floor joists 7′-3″.
3. Mark the floor joist layout on the rim joists following the framing plan. See Figure 6.

Figure 6
Completed layout on rim joists

4. Assemble the floor frame with 16d galvanized common nails. Be sure to check each joist for crown and install it with the crowned edge up.
5. Arrange the skids on a flat and level surface at the distance specified in the floor framing plans. See Figure 7.

Figure 7
Figure 7 illustrates step 5

Figure 8
Figure 8 illustrates step 7

6. Position the floor frame on top of the skids and measure the diagonals of the floor frame to make sure it is square. Adjust as necessary.
7. Install joist clip angles at each floor joist along the two outer skids using 1 ½″ joist hanger nails. See Figure 8.
8. Toe-nail each joist to the center skid with 16d galvanized nails.
9. Cut two pieces of plywood floor sheathing 4′-11″ long.
10. Install the first piece of plywood sheathing so that the edge is flush with an end floor joist and one end is flush with a rear rim joist. Use 8d galvanized box nails driven every 6″ along the edges and every 1′-0″ in the field.
11. Rip the second piece of plywood sheathing to fit.
12. Install the second piece of plywood sheathing flush with the rear rim joist and the opposite end joist. See Figure 9. Use 8d galvanized box nails driven every 6″ along the edges and every 1′-0″ in the field.
13. Cut six 1 x 6 x 8′-2″ decking boards.
14. Position the first board along the front edge of the floor front so it overhangs the rim joist and both end floor joists by 1 ½″. Fasten the board with 2″ deck screws.
15. Install the rest of the deck boards with 2″ deck screws. The last board will probably have to be ripped to fit. An example of a completed deck is shown in Figure 10.

Figure 9
Figure 9 illustrates step 12

Figure 10
Completed deck

Build the Walls

Once the floor frame has been completed and decking boards have been placed, you can start work on the walls of the playhouse. Wall construction begins by snapping lines for the location of the walls on the front, back, and sides of the plywood floor using a chalk box. After cutting wall plates and studs to the necessary sizes, you will then assemble, raise, and brace the walls. After the walls are raised they are sheathed with plywood.

Figure 11
Figure 11 illustrates step 1

PROCEDURE

1. Measure in 3 ½″ from all four sides of the plywood floor and snap chalk lines on the plywood floor for the wall plates. See Figure 11.
2. Cut four 2 x 4 x 4′-11″ side wall plates and four 2 x 4 x 7′-4″ front and rear wall plates.
3. Mark the stud and opening layouts on the plates following the floor plan. See Figure 12.

...snap chalk lines on the plywood floor...

Figure 12
Figure 12 illustrates step 3

Figure 13
Figure 13 illustrates step 10

4. Cut two 2 x 4 x 4′-4″ double top plates for the side walls.
5. Cut two 2 x 4 x 7′-11″ double top plates for the front and rear walls.
6. Cut twenty six 2 x 4 x 4′-11 ½″ studs.
7. Cut six 2 x 4 x 4′-6″ trimmer studs for the window and door openings.
8. Cut four 2 x 4 x 2′-3 ½″ sills.
9. Cut six 2 x 4 x 23 ½″ cripple studs.
10. Build the window and door headers. See Figure 13.
11. Build four stud-block-stud corners for the side walls.
12. Assemble all four walls.
13. Raise the side wall with the window and temporarily brace it in place.
14. Raise the back wall and secure it to the side wall. See Figure 14.

Figure 14
Figure 14 illustrates step 14

Figure 15
Figure 15 illustrates step 15

15. Raise the remaining side wall and secure it to the back wall. See Figure 15.
16. Raise the front wall and secure it to the two side walls. See Figure 16(A). Be sure all four walls are braced, plumb, and square and the double top plates are nailed properly. See Figure 16(B).

(A)

Figure 16
Figure 16 illustrates step 16

(B)

Figure 17
Figure 17 illustrates step 17

17. Install ½″ plywood sheathing to the sides and back wall frames with 6d box nails. Install the plywood from the bottom of the floor frame to ¾″ down from the top of the double top plate. See Figure 17.
18. Install ½″ plywood sheathing to the front wall ¾″ down from the top of the top plate to the deck board with 6d box nails. See Figure 18 for an illustration of how the project should look so far.

Figure 18
Figure 18 illustrates step 18

Install Porch Post and Beam

This playhouse has a post and beam porch. Two posts will be cut from 4 x 4 cedar or treated lumber and positioned at the front corners of the porch. The posts will be secured to the floor decking with metal post bases and 16d galvanized nails. A 4 x 6 beam will be installed on top of the posts using a post/beam cap and nails.

PROCEDURE

1. Cut two 4 x 4 x 4′-¼″ posts.
2. Install the metal post bases 1 ½″ from the front edge and end of the decking using 6d galvanized box nails. See Figure 19.

Figure 19
Figure 19 illustrates step 2

3. Set the posts in the metal post bases.
4. Fasten the post bases to the post with 6d galvanized box nails. See Figure 20.

Figure 20
Figure 20 illustrates step 4

Temporary braces

Figure 21
Figure 21 illustrates step 5

5. Plumb the posts and install temporary cross braces to keep them in place. See Figure 21.
6. Install a post/beam cap on top of each post using 6d galvanized box nails. See Figure 22.

Figure 22
Figure 22 illustrates step 6

7. Cut the 4 x 6 beam 8′-11″ long.
8. Set the beam crown up in the post caps on top of the posts so that it overhangs the posts 6″ on each end.
9. Fasten the beam to the post caps with 6d galvanized box nails. See Figure 23.
10. Measure diagonally between the posts to ensure that the posts and beam are square and adjust as needed. Once square is determined, install a temporary brace. See Figure 24.

...fasten the beam to the post caps...

Figure 23
Figure 23 illustrates step 9

Figure 24
Figure 24 illustrates step 10

Frame the Roof

To frame the roof you will cut pattern rafters for the front and rear of the house and the porch using the rafter templates as a guide. Once cut, the first two rafters of each type should be test fit using a 2 x 6 spacer block to simulate the ridge board. After they have been successfully tested, the pattern rafters can be used as a guide for cutting additional rafters. You will then cut and notch a 2 x 6 x 8′-11″ ridge board and install it along with the front and rear house rafters.

Figure 25
Figure 25 illustrates step 2

PROCEDURE

1. Cut one pattern rafter for each rafter type—a back rafter with an overhang, a front rafter, and a porch rafter with an overhang on one side—following the dimensions in the rafter drawing.
2. Test fit the front and back rafters by using a 2 x 6 spacer block, in place of the ridge board. See Figure 25.

3. Test fit the porch rafter by positioning it on the porch beam and the front wall of the playhouse. See Figure 26.
4. Adjust the pattern rafters to fit if necessary.
5. Cut the remaining back, front, and porch rafters so that you have a total of seven of each type.
6. Cut the rafters for the gable overhang. Do this by cutting four more porch rafters full length without the upper birdsmouth, four more back rafters without the birdsmouth, and cut four more front rafters without the level cut on the bottom end. See Figure 27. You can use the same patterns made in step 1, just trace the marks you need to cut.

Figure 26
Figure 26 illustrates step 3

Figure 27
Rafters for gable overhang

7. Cut the 2 x 6 x 8′-11″ ridge board.
8. Mark the rafter layout onto the front and rear wall plates following the wall stud layout. See Figure 28.

Figure 28
Figure 28 illustrates step 8

Figure 29
Figure 29 illustrates step 10

Figure 30
Figure 30 illustrates step 11

9. Lay out the notches on the ends of the ridge as shown in the drawings.
10. Measure 6″ in from the end of the ridge board and duplicate the rafter layout in step 8 on both sides of the ridge board. See Figure 29.
11. Cut out the notches laid out in step 9. See Figure 30.
12. Lay out the rafters on the porch beam. These rafters will be offset 1 ½″ from the layout on the front and rear walls to allow for the thickness of the front rafters. See Figure 31.

. . . lay out the rafters on the porch beam. . .

Figure 31
Figure 31 illustrates step 12

13. Install the front and rear rafters and the ridge board with 16d common nails.
14. Install the porch rafters to the porch beam with 16d common nails and face nail the porch rafters to the house rafters with 16d common nails. See Figure 32.
15. Cut two 2 x 4 x 3′-3″ collar ties and miter the ends at 26°.
16. Position the collar ties against the rafter pairs that fall to either side of the center rafters. Make sure the collar ties are level and the ends are ½″ away from the edges of the rafters. Face nail the collar ties to the rafters with three 12d common nails at each end. See Figure 33.
17. Cut the gable end wall plates to reach from the ridge to the wall plates. Install them with their outside edges flush with the outer rafters. See Figure 34.

Figure 32
Figure 32 illustrates step 14

Figure 33
Figure 33 illustrates step 16

Figure 34
Figure 34 illustrates step 17

Build the Gable Overhangs

The overhangs require fly rafters which are already cut, but the intersection of the porch rafter with the front rafter must be cut to fit at the proper angle. One porch fly rafter and one front fly rafter will be test fit and modified to the proper angle. Then they will be used as a pattern to modify the additional rafters.

PROCEDURE

1. Modify the porch and front fly rafter to create a pattern.
2. Clamp or tack the front fly rafter against an outer front rafter. See Figure 35.

Front fly rafter

Figure 35
Figure 35 illustrates step 2

...clamp or tack the front fly rafter...

Figure 36
Figure 36 illustrates step 3

3. Set a porch fly rafter in position on the porch beam. Align the top edge of the porch fly rafter with the top edge of the front fly rafter and scribe a line on the front fly rafter. See Figure 36.
4. Cut along the line scribed on the front rafter in step 3.
5. Test fit the rafters and adjust as needed.
6. Use the patterns to make one more porch fly rafter and one more front fly rafter.
7. Nail one porch fly rafter to one front fly rafter. See Figure 37.

Ridge cut

Cut made in step 4

Birdsmouth

Figure 37
Figure 37 illustrates step 7

Figure 38
Figure 38 illustrates step 11

8. Test fit the fly rafter assembly from step 7.
9. Nail the remaining porch fly rafter to the other front fly rafter.
10. Modify the remaining front rafter by aligning the fly rafter assembly from step 8 with the front rafter.
11. Scribe the portion of the rafter that extends beyond the bottom edge of the fly rafter assembly. See Figure 38.
12. Cut the front rafter on the line scribed in step 11.
13. Trace and cut second front rafter to match the rafter cut in step 12.
14. Install ½" sheathing on the gable end. See Figure 39.

Figure 39
Figure 39 illustrates step 14

Figure 40
Figure 40 illustrates step 18

15. Cut three pieces of 1 x 4s to 8′-11″. Two of these pieces will be used as sub fascia and the remaining piece will be used as a temporary fly rafter support.
16. Lay out two of the 1 x 4s cut in step 15 to match the rafter layout on the ridge.
17. Lay out the remaining 1 x 4 to match the layout on the porch beam.
18. Install one of the 1 x 4s laid out in step 16 on the back rafters as sub fascia. See Figure 40.
19. Install the 1 x 4 laid out in step 17 on the porch rafters as sub fascia. See Figure 41.

Figure 41
Figure 41 illustrates step 19

Figure 42
Figure 42 illustrates step 20

20. Install the second 1 x 4 laid out in step 16 on top of the front rafter just above the intersection of the porch rafter. See Figure 42.
21. Nail the modified front rafters and back rafters to the gable ends, aligning at the ridge and top end of rafter. See Figure 43.

...nail the modified front rafters...

Figure 43
Figure 43 illustrates step 21

Figure 44
Figure 44 illustrates step 22

22. Secure the fly rafter assembly to the ridge, porch beam, and temporary spacer. See Figure 44.
23. Secure the back fly rafter to the ridge and sub fascia. See Figure 45.

Figure 45
Figure 45 illustrates step 23

Install the Sheathing and Fascia

The roof of the playhouse will consist of fascia and sheathing covered with building paper and weather-resistant shingles. Once the ½″ plywood sheathing is installed using 6d nails, the trim fascia is installed. It is important to make sure the sheathing joints are aligned with the rafter joints at the point where the house roof meets the porch roof.

Figure 46
Figure 46 illustrates step 1

PROCEDURE

1. Install the ½″ plywood sheathing using 6d box nails. Make sure that the ends of the sheathing are installed so that they bear on half of the rafter. See Figure 46.
2. Install the 1 x 6 fascia trim along the gable overhangs. Hold the fascia flush with the top of the sheathing and attach with 6d galvanized finish nails.
3. Install the fascia along the front and rear eaves so that the front outside edge is flush with the top of the sheathing. Use a square or straightedge to align the front edge of the fascia with the top edge of the sheathing.

Apply Roofing Materials

The roof of the playhouse will be covered with self sealing asphalt shingles. These will be applied over a layer of building paper that is placed over the sheathing. Before installing shingles, you must establish a roof centerline. This project calls for 240 lb. 12″ x 3′-0″ 3 tab self-sealing asphalt shingles, although other sizes and types of shingles could be used. The shingles will be nailed in place in overlapping courses using chalk lines as guides.

PROCEDURE

1. Attach metal drip edge along the front and rear eaves. See Figure 47.
2. Apply 15 pound building paper over the sheathing maintaining a 2″ overlap of the building paper seams. See Figure 48.

Figure 47
Figure 47 illustrates step 1

Figure 48
Figure 48 illustrates step 2

Figure 49
Figure 49 illustrates step 3

Figure 50
Figure 50 illustrates step 6

3. Add drip edge along the gable ends, over the paper. See Figure 49.
4. Determine the width and length of the asphalt shingles. The materials list calls for 12″ x 3′-0″ shingles.
5. Measure and make a mark at each gable end a distance of 11 ⅜″ up from the outside edge of the drip edge.
6. At the marks made in step 5, snap a chalk line from one gable end to the other. See Figure 50. This line is where you will start your first reversed course of shingles while allowing a ⅝″ shingle overhang along the drip edge.
7. Repeat step 3 on the opposite side of the roof and gable ends.
8. Locate the exact center of the roof and mark it with a chalk line up to the ridge from the edge of the drip edge.
9. Repeat step 8 for the opposite side of the gable roof. This will be your roof centerline.
10. Install a reverse course of shingles aligned with the chalk line made in step 6. The solid surface of the shingle will be installed along the entire length of the roof edge and the ends of the first two starting shingles will be butted against the centerline made in step 9. See Figure 51.
11. After placing and nailing the starter strip, locate the centerline of the roof from previous chalk line in step 9.

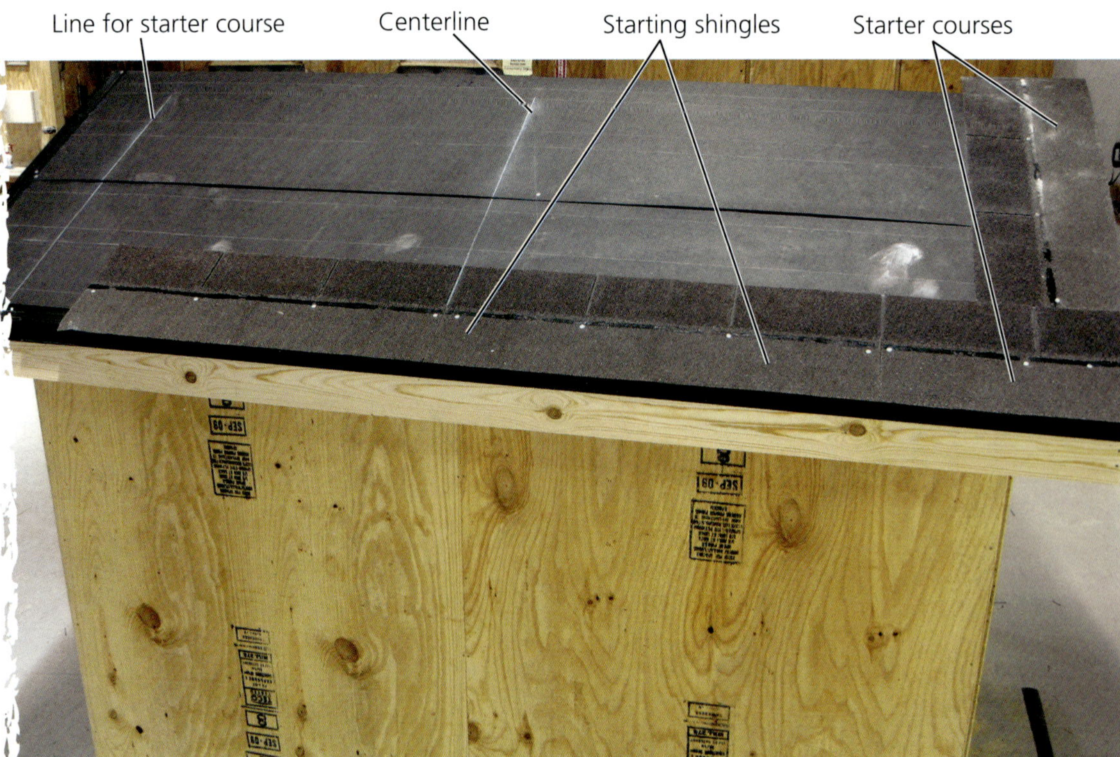

Figure 51
Figure 51 illustrates step 10

Figure 52
Figure 52 illustrates step 12

12. Center your first shingle on the chalk line directly on top of the starter strip, maintaining a ⅝″ overhang along the drip edge and nail the shingle in place. See Figure 52.
13. Continue to butt adjacent shingles left and right of the first shingle laid on top of starter strip in step 12. Cut the last shingles of the course to size using a utility knife while maintaining a ⅝″ overhang along the metal drip edge.
14. Make a mark at each end of the gable roof 5″ above top edge of the first course of shingles and snap a horizontal chalk line. This will be used to aid in the alignment of the second course of shingles and to set the shingle exposure.
15. Start the second course of shingles, again working from the center of the roof. Place the end of the shingle directly over the center of the shingle below. Make sure the shingle is aligned to the center chalk line made in step 10 and the top is aligned with the horizontal chalk line made in step 14. See Figure 53.

Figure 53
Figure 53 illustrates step 15

Figure 54
Shingled roof without ridge cap

16. Secure the shingle with four nails, one above each tab cutout.
17. Butt the next shingle against the first shingle aligned with the horizontal chalk line and nail in place with four nails.
18. Continue placing shingles out to the gable ends of the roof on each side of the centerline. Cut the end shingles to size using a utility knife while maintaining a ⅝″ overhang along the metal drip edge.
19. Repeat the process to the ridge of the roof for the remaining courses while making sure to offset the cutouts of the shingles with the previously installed course.
20. Repeat steps 9–18 to install the shingles for the other side of the roof. See Figure 54.
21. Cut one shingle into three 12″ shingles to begin the ridge installation. See Figure 55.

Figure 55
Figure 55 illustrates step 21

Figure 56
Figure 56 illustrates step 22

22. Overlap a 12″ shingle equally on each side of the ridge and install two nails, starting on either side of the ridge working in one direction. See Figure 56.
23. Install the remaining ridge shingles maintaining a 5″ shingle exposure until complete. There will be two exposed nails holding the last shingle in place. See Figure 57.

Figure 57
(a) Figure 57 illustrates step 23

(b) Completed roof

Student Name:______________________________ Date: ______________

Playhouse Project Evaluation

PROCEDURE	CRITERIA AS SPECIFIED BY THE PROJECT	POSSIBLE POINTS	SCORE
Floor System	Joist layout	5	
	Rim joists attachment	5	
	Plywood installation	5	
	Parallel sides equal	5	
	Floor length	5	
	Floor width	5	
	Floor system square	5	
	Fastener installation	5	
	Subtotal	*40*	
Wall System	Stud layout	5	
	Rough opening for door	5	
	Door opening frame	5	
	Rough opening for window	5	
	Window opening frame	5	
	Header location	5	
	Headers built correctly	5	
	Walls square	5	
	Walls flush at corners	5	
	Double plate installation	5	
	Fastener installation	5	
	Subtotal	*55*	
Porch System	Post installation	5	
	Beam installation	5	
	Subtotal	*10*	
Roof System	Rafter layout	5	
	Rafter pitch correct	5	
	Plumb cuts	5	
	Birdsmouth cuts	5	
	Gable overhang correct	5	
	Ridge board installation	5	
	Gable stud installation	5	
	Sheathing installation	5	
	Fascia installation	5	
	Metal drip edge installation	5	
	Asphalt shingle installation	5	
	Fastener installation	5	
	Subtotal	*60*	

Playhouse Project Evaluation *(continued)*

PROCEDURE	CRITERIA AS SPECIFIED BY THE PROJECT	POSSIBLE POINTS	SCORE
Overall Dimensions	Length correct	5	
	Width correct	5	
	Wall height correct	5	
	Subtotal	*15*	
General	Proper tool handling	5	
	Followed direction	5	
	Cleaned up	5	
	Safe work practices	5	
	Subtotal	*20*	
		Student score:	

Total Possible Points = 200

Suggested minimum acceptable score: 140 points or ______

Student's Signature: ______________________ Teacher's Signature: ______________________